CAMBRIDGE LIBRARY COLLECTION

Books of enduring scholarly value

Earth Sciences

In the nineteenth century, geology emerged as a distinct academic discipline. It pointed the way towards the theory of evolution, as scientists including Gideon Mantell, Adam Sedgwick, Charles Lyell and Roderick Murchison began to use the evidence of minerals, rock formations and fossils to demonstrate that the earth was older by millions of years than the conventional, Bible-based wisdom had supposed. They argued convincingly that the climate, flora and fauna of the distant past could be deduced from geological evidence. Volcanic activity, the formation of mountains, and the action of glaciers and rivers, tides and ocean currents also became better understood. This series includes landmark publications by pioneers of the modern earth sciences, who advanced the scientific understanding of our planet and the processes by which it is constantly re-shaped.

A Geognostical Essay on the Superposition of Rocks in Both Hemispheres

The explorer and multi-disciplinary scientist Alexander von Humboldt (1769–1859) was a prominent figure in the European scientific community of the eighteenth and nineteenth centuries and the first to make a scientific survey of South and Central America. His travels alone brought him widespread recognition, but the extensive field notes and research he undertook were developed further on his return. Originally published in French and translated in 1823, this work brought his geological speculations to a British audience. Humboldt explores the positioning of different types of rocks across the globe, and the causes behind these formations. He also hypothesises that the flora of these areas are affected by the geology, which in turn is influenced by the thermal currents of the earth's molten core. These insights into rock formations are also key to Humboldt's theory of continental drift, now recognised as resulting from the shifting of the continental plates.

Cambridge University Press has long been a pioneer in the reissuing of out-of-print titles from its own backlist, producing digital reprints of books that are still sought after by scholars and students but could not be reprinted economically using traditional technology. The Cambridge Library Collection extends this activity to a wider range of books which are still of importance to researchers and professionals, either for the source material they contain, or as landmarks in the history of their academic discipline.

Drawing from the world-renowned collections in the Cambridge University Library and other partner libraries, and guided by the advice of experts in each subject area, Cambridge University Press is using state-of-the-art scanning machines in its own Printing House to capture the content of each book selected for inclusion. The files are processed to give a consistently clear, crisp image, and the books finished to the high quality standard for which the Press is recognised around the world. The latest print-on-demand technology ensures that the books will remain available indefinitely, and that orders for single or multiple copies can quickly be supplied.

The Cambridge Library Collection brings back to life books of enduring scholarly value (including out-of-copyright works originally issued by other publishers) across a wide range of disciplines in the humanities and social sciences and in science and technology.

A Geognostical Essay on the Superposition of Rocks in Both Hemispheres

Alexander von Humboldt

CAMBRIDGE UNIVERSITY PRESS

Cambridge, New York, Melbourne, Madrid, Cape Town,
Singapore, São Paolo, Delhi, Mexico City

Published in the United States of America by Cambridge University Press, New York

www.cambridge.org
Information on this title: www.cambridge.org/9781108049498

© in this compilation Cambridge University Press 2012

This edition first published 1823
This digitally printed version 2012

ISBN 978-1-108-04949-8 Paperback

This book reproduces the text of the original edition. The content and language reflect the beliefs, practices and terminology of their time, and have not been updated.

Cambridge University Press wishes to make clear that the book, unless originally published by Cambridge, is not being republished by, in association or collaboration with, or with the endorsement or approval of, the original publisher or its successors in title.

A

GEOGNOSTICAL ESSAY

ON THE

SUPERPOSITION OF ROCKS,

IN BOTH HEMISPHERES.

BY

ALEXANDRE DE HUMBOLDT.

TRANSLATED FROM THE ORIGINAL FRENCH.

LONDON:

PRINTED FOR

LONGMAN, HURST, REES, ORME, BROWN, AND GREEN,

PATERNOSTER-ROW.

1823.

PREFACE.

THE work which I now submit to the judgment of geognosts includes nearly the whole of the subject of positive geognosy. If I have accomplished the end which I proposed, the phenomena of the most remarkable superpositions of rocks in both hemispheres, north and south of the equator, will appear arranged in the order of their mutual relations. I cannot flatter myself with having succeeded in comprising within so narrow a space, so great a variety of objects; but I hope that my work will contain two sources of interest; that of making known a considerable number of observations which had not hitherto been published, and that of presenting some general views on the succession of those rocks which have been considered as the *terms* of a simple or a periodical series.

The comparison of the rocks of the old world with those of the Cordillera of the Andes, has been deduced exclusively from my own researches. To guard myself against the danger of first impressions, and the errors which might arise from certain prejudices, I have, within a few months past, read over all the manuscripts which I had written during my travels; and I have compared the descriptions with the sections and profiles of the mountains which were drawn on the spot. After having considered the whole of their geognostic relations, I have confined myself to those which appeared to me the most certain or the most probable, and I frankly state what still demands a more particular examination.

Previous to the application of systematic names to the formations of the Andes, of the Oronoco, of the Amazon, or of New Spain, I have described their various relations of position, of composition, and of structure. This method, which I have constantly followed, will enable the reader to decide more easily on the degree of confidence which my arrangements merit. If it be recollected, that before my travels in equinoctial America, scarcely any rock in that country had been named, and that I could not be guided in the study of *superpositions* by any anterior observations, it will, I hope, appear less surprising, should all my descriptions not be found equally perfect. The articles which I have devoted to the different formations are of unequal length, according to the number of facts which I have been able to state respecting them.

In this *geognostical essay*, as well as in my researches on the *isothermal lines*, on the *geography of plants*, and on the laws which have been observed in the *distribution of organic bodies*, I have endeavoured, at the same time that I presented the detail of the phenomena, to generalize the ideas respecting them, and to connect them with the great questions in natural philosophy. I have dwelt chiefly on the phenomena of *alternation*, of *oscillation*, and of *local suppression*, and in those which result from the *passage* of one formation to another in consequence of *interior developement*. These subjects are not mere theoretical speculations; far from being useless, they lead us to the knowledge of the laws of nature. It would degrade the sciences to make their progress depend solely on the accumulation and study of particular phenomena.

It is already many years since I first announced the table of positions which I now publish. The hesitation with which it is usual to proceed to the printing of a work long expected, would perhaps have still farther retarded this publication, had I not been compelled to it by the duties of friendship. M. Levrault, rector of the Academy of Strasburg, one of those estimable and useful men who, while existing, receive from their cotemporaries the tribute of gratitude which they merit, requested my co-operation in the grand literary work which he had confided to the celebrated professors of the Museum of Natural History of Paris; and he succeeded in overcoming the repugnance which I have always felt for engaging in this kind of labour. I promised him that I would undertake for his "Dictionnaire des Sciences Naturelles," the article *Geography of Plants*. Some unforeseen occupations having prevented me from fulfilling my promise, this article has been supplied by M. De Candolle, with the distinguished talent that characterises all his works; and I have only added to it researches on the numerical relations of vegetable forms, and on the distribution of those forms in the different climates. As a kind of compensation, I offered to write the article *Geognosy*, in which would be comprehended the description of the several formations. The following work consists of this article, which is now printed separately. It is nearly of the same extent as the article *formation* (*terrain*), which an excellent geognost, M. de Bon nard, has given in the "Dictionnaire d'Histoire Naturelle," which is less voluminous, and is published by M. Deterville. It appeared to me that one could not better arrange the facts according to their natural relations, than by devoting forty sections to forty *independent formations.*

I have been particularly careful to indicate the localities of the most interesting phenomena of position; and I have frequently added to them the results of my barometrical measurements. When there has been any doubt with respect to those countries of which we have only very imperfect maps, I have mentioned such latitudes as I had determined during my excursions in the Cordilleras.

I have explained, at the end of the work, the principles of a *geognostic pasigraphy*; and have wished to shew that by means of a very simple notation, and by omitting the structure and composition of rocks, we may express with great facility the most complicated relations that exist between the position and the periodical recurrence of formations. This method of notation and concise language render evident the identity of phenomena, which, when disguised by accidental circumstances, might at first appear to be very different. The *pasigraphic notation, which proceeds by series,* and which presents an almost *algorithmic* method, is more susceptible of perfection than the *imitative* or figured *pasigraphy.* Both these appear to me to be important in geology; for it is with the pasigraphic language as with languages in general; the ideas become more clear in proportion as the signs which express them are improved.

INTRODUCTION.

In geognosy, the word *formation* either denotes the manner in which a rock has been produced, or it designates an assemblage of mineral masses so intimately connected, that it is supposed they were formed at the same epoch, and that they present, in the most distant parts of the earth, the same general relations, both of composition, and of situation with respect to each other. Thus the *formation* of obsidian and of basalt is attributed to subterraneous fires; and it is also said that the *formation* of transition clay-slate contains Lydian stone, chiastolite, ampelite, and alternating beds of black limestone, and of porphyry. The first acceptation of the word is the most conformable to the genius of the French language; but it relates to the origin of things, and to an uncertain science founded on geogonic hypotheses. The second acceptation, now generally received by the French mineralogists, has been borrowed from the celebrated school of Werner, and indicates, not what is supposed to have been, but what now exists.

In the geognostic description of the globe, we may distinguish different modes of grouping mineral substances, as we ascend to more general ideas. Rocks, which alternate with each other, which are found usually together, and which display the same relations of position, constitute the same *formation;* the union of several formations constitutes a geological series or a district (*terrain*) ; but the terms rocks, formations, and *terrains* *, are used as synonymous in many works on geognosy.

The diversity of rocks, and the relative position of the beds which form the oxidated crust of the globe, have, from the most remote times, fixed the attention of men. Wherever the working of a mine was directed on a mass of salt, coal, or clay iron-stone, covered by successive beds of a different nature, it gave rise to ideas more or less precise, on the arrangement of the rocks peculiar to a formation of small extent. Possessed of this local knowledge, but influenced by prejudices having their source in habit, miners spread themselves over contiguous countries, and, as geognosts have often done in our days, they decided upon the positions of rocks, of the nature of which they were ignorant, according to incomplete analogies, and the confined ideas they had originally acquired. These errors must have had a fatal influence on the suc-

* We have no word in the English language that will accurately express *terrain*, as used in geology by the French: it here means a series of formations: but sometimes also *terrain* denotes a tract or district consisting of a particular class of rocks.—*Translator.*

cess of their researches. Instead of studying the connection between two successive formations, instead of extending the first types of formations which had been impressed on their minds, they imagined that each portion of the globe differed in its geological constitution. This very ancient popular opinion has been adopted and maintained by very distinguished men in different countries; but when geognosy was raised to the rank of a science, when the art of interrogating nature was improved, and when journies to distant countries furnished a more exact comparison between different formations, great and immutable laws were recognised in the structure of the globe, and in the superposition of rocks. The most striking analogies in the position, composition, and the included organic remains, of contemporary beds, were then observed in both hemispheres; and in in proportion as we consider *formations* under a more general point of view, their *identity* daily becomes more probable.

In fact, when we examine the solid mass of our planet, we soon perceive that some of the substances, which oryctognosy (or descriptive mineralogy) has made known to us separately, are found in *constant associations*, and that those associations, which are called compound rocks, do not vary, like organised beings, according to the difference of latitude, or of the isothermal bands under which they are placed. Geognosts, who have travelled through the most distant countries, have not only found, for the most part, in the two hemispheres, the same simple substances, quartz, feldspar, mica, garnet, and hornblende; but they have also observed that mountain-masses display every where the same rocks; that is, the same assemblages of mica, quartz, and feldspar, in granite; of mica, quartz, and garnets, in mica-slate; and of feldspar and hornblende, in syenite. If it has sometimes been considered, that a rock belongs exclusively to a single portion of the globe, subsequent researches have shown, that it also occurs in regions the most distant from its first locality. Thus we are almost led to admit, that the formation of rocks has been independent of the diversity of climates, and perhaps anterior to its existence.* There is an identity even in those rocks where organized bodies are the most variously modified.

But this identity of composition, this analogy which is observed in the association of certain simple mineral substances, may be independent of the analogy of their position, and their succession. Specimens of the same rocks that are found in Europe may have been brought from the islands of the Pacific Ocean, or the Cordilleras of the Andes; yet perhaps we are not authorised to conclude from thence, that these rocks are superposed in a similar order, and that from the discovery of one, it can be predicted, with certainty, what the others are which occur in the same places. Geognosts, who are

* Humboldt, Geography of Plants, 1807, p. 115. Idem, Views of the Cordilleras, vol. i. p. 122.

devoted to the study of the laws of unorganized nature, should direct their labours towards the recognition of these analogies of respective position. An attempt is made in the following table, to collect what is known with most certainty of the superposition of rocks in both continents north and south of the equator. These *types of formations* will not only be extended, but also variously modified, in proportion to the increase of travellers practised in geognostic observations, and also when complete monographies of different countries far remote from each other shall furnish more precise results.

The developement of the order which is found to exist in the superposition of rocks forms the most important part of geognostic science. It must be allowed, that great difficulties often present themselves in the observation of positions; either when we cannot arrive at the junction of two adjoining formations, or when they do not exhibit a regular stratification, or when their position is not *uniform*, that is, when the strata of the superior formation are not parallel to those of the lower. But these difficulties (and it is one of the chief advantages in observations which extend to a considerable portion of our planet) diminish, or even disappear entirely, on comparing together several formations of great extent. The order of superposition and the relative age of rocks are facts susceptible of being determined, like the structure of a plant, the proportion of the elements in a chemical compound, or the height of a mountain above the level of the sea. True geognosy describes the exterior crust of our globe such as it exists in our days. This science has no less certainty than the physical descriptive sciences in general: on the contrary, whatever relates to the ancient state of our planet, to those fluids which, it is said, held all mineral substances in solution, to those seas which have covered the summit of the Cordilleras, and have afterwards disappeared, is as uncertain as the formation of the atmosphere of the planets, as the various migrations of plants, or the origin of the different varieties of our species; yet the time is still not very remote when geolists were occupied from choice in the solution of these problems, and with this fabulous period of the physical history of the earth.

To render more intelligible the principle upon which this tabular arrangement of the superposition of rocks is constructed, it ought to be preceded by some remarks deduced from the practical study of different formations. We shall begin by observing, that it is not easy to circumscribe the limits of a formation. The limestone of the Jura and the Alpine limestone, entirely distinct in one region, appear sometimes closely connected in another. What proves *the independence of a formation*, as M. de Buch has well observed, is its immediate superposition on rocks of a different nature, which, consequently, ought to be considered as more ancient. The red sandstone constitutes an independent formation, because it is superposed indifferently on black transition limestone, mica-slate, and primitive granite; but in a region where the great formation of syenite and porphyry predominates, those two rocks constantly alternate. It thence results, that the syenite is subordinate to the porphyry, and scarcely any where covers, by itself, the clay-slate of transition or the primitive gneiss. But the independence of formation in no manner excludes *uniformity* or *concordance of position*; it rather excludes the oryctognostic passage between two superposed formations. Transition formations have very often the same direction and the same dip as primitive formations; and yet, however near may be the dates of their origins, we are not the less justified in considering the anthracitous mica-slate, or grauwacke alternating with porphyry, as two formations independent of the granite and the primitive gneiss which they cover. The uniformity of position (*gleïchformigkeit der lagerung*) furnishes no argument against the independence of formations, or the considering a rock as a distinct formation. It is because the independent formations are placed indifferently on all the most ancient rocks, (chalk upon granite, or red sandstone on primitive mica-slate,) that the union of a great number of observations made on very remote points becomes eminently useful in determining the relative age of rocks. In order to ascertain if the zircon syenite be a transition rock, we must shew that it is placed on formations posterior to the black limestones containing orthoceratites. The observations made by M. Beudant, one of the most distinguished geologists of the present day, on the porphyries and syenites of Hungary, may throw great light on the formation of the American Andes: and thus, a plant discovered in India may point out the natural affinity between two families of plants of equinoxial America.

The order which is followed in the *tabular arrangement of formations*, is, that of the place, and the respective position of the rocks. I do not pretend that this situation and position has been actually observed in every region of the earth; I only state them, such as they appeared to me the most probable; after having compared a great number of facts that I have collected. I have been guided by the idea of the relative age of rocks, in this yet very imperfect labour which was begun long before my voyage to the Cordilleras of the new continent, in the year 1792, when, upon leaving the school of Freyberg, I was appointed to the direction of the mines in the Fichtelgebirge.

A rock may vary in composition: some of its integrant parts may be subtracted from it, or other substances may be found disseminated in it; and yet, in the opinion of a geognost who has studied the superposition of formations, the rock ought not to change its denomination. Under the equator, as in the north of Europe, the beds of a real transition syenite lose their hornblende without becoming another rock. The granites on the banks of the

Oronoco sometimes contain hornblende, and yet should still be considered as primitive, although they are not of the first or most ancient formation. These facts have been admitted by all experienced geognosts. The essential character of the identity of an independent formation is its relative position, or the place which it occupies in the general series of formations. (Vide the classical memoir by M. de Buch, *Heber der Begriff einer gebirgsart*, in the Mag. der Naturf. 1810. p. 128—133.) For the same reason a mere solitary fragment, an insulated specimen of a rock found in a collection, cannot be *geognostically* determined, that is, as belonging to one of the numerous beds of which the crust of our planet is composed. Chiastolite, the accumulation of carbon, or the nodules of compact limestone in clay-slate, nigrine or epidote in syenites (alternating with granites or porphyries), and conglomerates contained in anthracitous mica-slate, point out transition formations; in the same manner as from the important labours of M. Brongniart, petrifactions of shells preserved entire indicate with precision certain beds of the tertiary formation. But these observations, where we are guided by disseminated substances, or by characters simply zoological, comprehend only a small number of rocks of late origin; and observations of this kind often lead only to negative facts. Characters drawn from the colour of the grain, or the small veins of carbonate of lime that run through calcareous rocks; those that are derived from the fissile nature, or the silky lustre of the clay-slate; the general aspect, and the wavy character of the scales of mica in mica-slates; the size and the colour of the crystals of feldspar in granites of various formations; all these circumstances may, like every thing connected with the *habitus* of minerals, lead the most acute observer into error. No doubt black and white are the distinguishing colours of the primitive and transition limestones; no doubt the formation of the Jura, particularly in the superior part, is generally divided into thin beds that are whitish, with a fracture dull, even, or nearly flat conchoidal; but, in mountains of transition limestone, there exist insulated masses, which in their colour and texture resemble in their oryctognostic characters the Jura formation; there are also hills of the tertiary formation on the south of the Alps, where rocks, subsequent to the chalk, and resembling the limestone used for lithography, are found analogous to the fissile and dull limestone of the Jura. If we prefer giving to formations names derived from their oryctognostic characters only, the various strata of the same compound rock, when its thickness is considerable, and when it can be traced far in the line of its direction (streichungslinie), may appear often to belong to different rocks, according to the points from which specimens are taken; consequently we can scarcely determine any thing geognostically in collections, but the *suites of rocks*, of which the mutual superposition is known.

In advancing these opinions on the sense which

we ought to attach to the term *independent formations*, as it relates to the following tabular arrangement of their position, I am far from overlooking the eminent services, which the most detailed oryctognostic examination, and the profound study of the composition of rocks, have rendered to modern geognosy, and especially to the science of the position, and respective situation of rocks; although, according to the important discoveries of M. Hauy on the intimate nature of inorganic and crystallized substances, there cannot exist, properly speaking, a *passage* from one mineral substance to another; (Cordier, *sur les Roches Volcaniques*, p. 33.; and Berzelius, *Nouveau Système de Minéralogie.*) The passage of the *base*, or *mass of rocks*, is not confined to those formations that are generally distinguished by the name of compound rocks. Those considered as simple, such as transition limestone, or secondary limestone, are in part amorphous varieties of mineral species of which there exists a crystallized type; and partly aggregates of clay, carbon, &c., which cannot be accurately determined. It is on the variable proportions of these heterogeneous mixtures that the passages of marly limestones to other schistose formations are founded; (Hauy, *Tableau Comparatif de la Cristallographie*, p. 27. 30.) All the amorphous bases of rocks, however homogeneous they may appear at the first aspect, the bases of porphyries and euphotides (serpentine), as well as those black problematic masses that constitute the *basanite* (basaltes) of the ancients, and which are not all greenstone overcharged with hornblende, are capable of being submitted to a mechanical analysis. M. Cordier has employed that analysis in the most ingenious manner to greenstones, dolerites, and other volcanic productions more recent. The most minute oryctognostic examination cannot be unimportant to the geognost who wishes to determine the relative age of formations. It is by this kind of examination that we obtain a just idea of the progressive manner in which, by *interior developement,* (that is, by a very slow change in the proportions of elements of the mass), the passage takes place from one rock to another neighbouring rock. The schists of transition, of which the structure seems, at first sight, so different from that of porphyry or granite, present to the attentive observer striking examples of insensible passages to rocks that are granular, porphyritic, or granitoïd. At first these schists become greenish, and harder; in proportion as the amorphous paste acquires hornblende, it passes to those amphibolic trap rocks that were formerly mistaken for basalt. In other places, the mica, concealed at first in the amorphous mass, is developed, and separates into plates distinctly crystallized; at the same time, the feldspar and quartz become visible, and the mass assumes a granular appearance, with elongated grains; this is a true transition gneiss. The grains lose by degrees their common direction; the crystals are grouped around several centers, and the rock becomes a granite, or a transition syenite. In other cases the quartz alone is developed, augments, becomes formed into round nodules, and the schist passes to a grauwacke very distinctly characterized. By these certain signs, geognosts who have long studied nature, never fail to recognise the proximity of granular, granitoïd, or arenaceous rocks. Analogous passages between primitive mica-slate and a porphyritic rock, and the return of this last to gneiss, are observed in the eastern parts of Switzerland; (Vide the luminous remarks of M. de Raumer, *Fragmente*, p. 10. 47.; M. Leopold de Buch, in his *Voyage de Glaris à Chiavenna*, 1803, inserted in the *Mag. der Berl. Naturf.* vol. i. p. 119.) But those passages are not always insensible and progressive; rocks often succeed each other abruptly, and in a distinctly separate manner. Often, (for example, in Mexico, between Guanaxuato and Ovexaras), the limits between the schists, the porphyries, and the syenites, are as distinct as between the porphyries and lime-stones; but, even in this case, heterogeneous interposed beds indicate geognostical relations with the superposed rocks. Thus transition granite, in the syenitic formation, presents beds of trap, because it becomes charged with hornblende; and in a similar manner these same granites pass sometimes to euphotide; (Buch, *Voyage en Norwège*, tom. i. p. 138., tom. ii. p. 83.)

From these considerations it follows, that the mechanical analysis of amorphous masses, by means of slight triturations, and washing, (a method of analysis of which M. Fleuriau de Bellevue made the first successful essay), throws light, first, on the large crystals, which are insulated, and which separate from the microscopic crystals in the mass; secondly, on the mutual passages of certain rocks superposed on each other; thirdly, on the subordinate beds which are of the same nature as one of the elements of the amorphous mass. All these phenomena are produced, we may say, by internal developement, and through a variation sometimes slow, sometimes very sudden in the constituent parts of a heterogeneous mass. Crystalline molecules, invisible to the eye, are found increased and disengaged from the close texture of the mass; and by degrees they become, by grouping and mixture with other substances, interposed beds of considerable thickness: sometimes they even become new rocks.

The interposed beds merit, above all, the deepest attention (Leonhard, Kopp, and Gœrtner, *Propæd. der Miner*, p. 158.) When two formations succeed each other immediately, it happens that beds of the one begin at first to alternate with beds of the other, until, (after these preludes to a great change,) a new formation appears without any subordinate beds; (Buch, *Geogn. Beob.* tom. i. p. 104—156.; Humboldt, *Relation Historique*, tom. ii. p. 140.) The progressive developement of the elements of a rock may consequently have a marked influence on the respective position of mineral masses. Their effects belong to the domain of geognosy; but to discover and appreciate them, the observer must have recourse to the profound knowledge of oryctognosy, and above all, to that of modern crystallography.

In examining the intimate relations by which we often see the phenomena of composition connected with those of position, I had no intention of speaking of the purely oryctognostical method, that which considers rocks solely according to the analogy of their composition; (*Journal des Mines*, tom. xxxiv. No. 199.) This is a true classification, from which every idea of superposition is abstracted, but which may not the less give rise to interesting considerations on the constant grouping of certain minerals. A method wholly oryctognostical multiplies the name of rocks more than the wants of geognosy require, when positions only are considered. According to the changes which mixed rocks undergo, the same bed of great extent and thickness may contain parts (we must again observe) to which the oryctognost, classing rocks according to their composition, would give denominations entirely different. These remarks have not escaped the learned author of the *Classification Minéralogique des Roches;* they could not but occur to an experienced geognost, who has so well investigated the superpositions of the rocks which he has examined. "We must not confound," says M. Brongniart, in his late memoir on the position of the Ophiolites,

" the respective situations, the order of the superposition of formations, and the rocks of which they are composed, with descriptions that are simply mineralogical (oryctognostical); their confusion would throw disorder into the science, and retard its progress." The tabular arrangement which we give at the end of this view, is, in no respect, what may be called a classification of rocks; nor have we collected, under the title of distinct sections, (as in the ancient geognostical method of Werner, or in the excellent *Traité de Geognosie* of M. D'Aubuisson,) all the primitive formations of granite, and all the secondary formations of sandtone and limestone. We have endeavoured, on the contrary, to place each rock as it is found in nature, according to the order of its superposition, or its respective age. The different formations of granite are separated by gneiss, mica-slate, black limestones, (of transition), and grauwackes. In the rocks of transition, we have removed the porphyry and syenite of Mexico and Peru, that are anterior to the grauwacke, and to the limestone with orthoceratites, from the much newer formation of porphyry and zircon syenite of Scandinavia. In the secondary rocks, the sandstone with oolite of Nebra, which is posterior to the alpine limestone or zechstein, is removed from the red sandstone (coal sandstone), which belongs to the same formation as the secondary porphyry and mandelstein. According to the principle which we have adopted, the same names of rocks are found several times in the same table. The anthracitous mica-slate (in the transition formation), is separated, by a great number of older formations, from the mica-slate anterior to the primitive clay-slate.

Instead of a *classification* of the granitic, schistose, calcareous, and arenaceous aggregated rocks, I wished to present a sketch of the geognostical structure of the globe, or a table in which the superposed rocks succeed each other from below upwards, as in the ideal sections which I drew in 1804, for the use of the *School of Mines of Mexico*, and of which many copies have been circulated since my return to Europe. (*Bosquejo de una Pasigrafia geognostica, con tablas que enseñan la estratificacion y el parallelismo de las rocas en ambos continentes, para el uso del Real Seminario de Mineria de Mexico.*) These pasigraphic pictures exhibited, together with my own observations, whatever had been collected with most precision at that period, on the position of primitive, intermediary, and secondary rocks in the antient continent. They presented, with the type which may be regarded as the most general, the secondary types, or the beds which I have named *parallels.* This method has been followed in the present sketch, which I now publish. My *parallel* formations are *geognostic equivalents;* they are those rocks that represent each other, (*Traité de Géologie* by M. d'Aubuisson, tom. ii. p. 255.) In England, and on the continent of

Europe opposite, there does not exist an identity in all the formations; there exist only *equivalents* or *parallel formations*. The situation of our coal, situated between the transition rocks and the red sandstone, that of the rock-salt, which is found on the continent in the alpine limestone, (zechstein), the position of our oolites in the sandstone of Nebra and in the Jura limestone, may guide the geognost in the comparison of distant formations. Coal measures are observed in England placed on transition formations; as, for example, on the mountain limestone of Derbyshire and South Wales, or on the sandstone of transition, the old red sandstone of Herefordshire. I thought I recognized in the magnesian limestone, the red marl, lias, and white oolites of Bath, the *united formations* of alpine limestone with rock-salt, of sandstone with oolites, (bunte sandstein), and of Jura limestone. In comparing the formations of more or less distant countries, those of England and France, of Hungary and Mexico, of the secondary basin of Santa Fé de Bogota and Thuringia, we must not expect to find for each rock a parallel rock; we must recollect that one single formation *may represent* many others. Thus the beds of clay inferior to the chalk may, in France, (Cap La Hève, Vaches Noires, near Caen), be separated in the clearest manner from the beds of oolitic limestone; while in Switzerland, Germany, and South America, they have for *equivalents* beds of marl subordinate to the Jura limestone. Gypsum, which in one district forms only subordinate beds in the alpine limestone, or in the oolitic sandstone, assumes, in another district, all the appearance of an independent formation, and is found placed between the alpine limestone, and the oolitic sandstone, between this sandstone and the muschelkalk, (limestone of Gœttingue). The learned professor of Oxford, M. Buckland, whose extensive researches have been alike useful to the geognosts of England and of the Continent, has recently published a table of *parallel* formations, or, as he calls them also, *equivalents of rocks,* which extends but from the forty-fourth to the fifty-fourth degree of northern latitude, and which merits the greatest attention; (*On the Structure of the Alps and their Relation with the Rocks of England,* 1821.)

As in the history of ancient nations, it is easier to verify the series of events in each country, than to determine their mutual coincidence, so also, we easier attain the most exact knowledge of the superposition of formations in insulated regions, than we can determine the relative age, or the parallelism of formations that belong to different systems of rocks. Even in countries little distant from each other, as in France, Switzerland, and Germany, it is not easy to fix the relative antiquity of the muschelkalk, of the molasse of Argovia, and of the quadersandstein of the Hartz, because the widely distributed rocks are often wanting which might serve, according to the happy expression of M. de Gruner, for a *geognostical horizon*, and with which the three formations we have just mentioned may be compared. When rocks are not in immediate contact, we can judge of their *parallelism* only by their age, as relative to the other formations by which they are united.

These researches of *comparative geognosy* will long exercise the sagacity of observers; and it is not surprising that those who have been disappointed in finding every formation, with all the circumstances of its position, interior structure, and subordinate beds, should end by denying all analogy of superposition. I enjoyed the advantage of visiting, before my voyage to the equator, a great part of Germany, France, Switzerland, England, Italy, Poland, and Spain. My attention was particularly directed, during my travels, to the *position* and *succession* of formations, a class of phenomena which I purposed treating of in a separate work. In South America, whilst traversing in various directions the vast tract which stretches from the chain on the coast of Venezuela to the basin of the Amazon, I was struck by the conformity of superposition exhibited in the two continents (vide my first *Sketch of a Geological View of Equinoctial America*, in the *Journal de Physique*, tom. liii. p. 30.) Posterior observations, comprehending those in the Cordilleras of Mexico, New Grenada, Quito, and Peru, from the twenty-first degree of north latitude to the twelfth degree of south latitude, confirmed those first ideas. The types of formations appeared to me to be rather enlarged than altered essentially. But in speaking of the analogies which have been observed in the positions of rocks, and the uniformity of those laws which exhibit to us the general order of nature, I can produce higher testimony than my own, that of the geognost whose labours most of all have advanced the knowledge of the structure of the globe. M. Léopold de Buch has extended his researches from the Canary Islands to beyond the Polar Circle, as far as the seventy-first degree of latitude. He has discovered new formations placed between formations formerly known; and in the primitive formations as well as in those of transition, in the secondary as well as in the volcanic rocks, he was struck with the great analogical features that characterise the aspect of formations, even in the most distant regions.

We must distinguish from that scepticism which denies all order in the position of rocks, an opinion renewed from time to time among the most experienced observers, according to which the formation of granite-gneiss, of grauwacke, of alpine limestone, and of chalk, uniformly superposed in different countries, seldom correspond among themselves in the age of the homonymous elements of each series. It has been thought that a secondary rock may have been formed on one point of the globe, at a period when the transition rocks did not yet exist in another point. This supposition does not include those granitic rocks that cover limestones filled with orthoceratites and consequently posterior to the primitive rocks. It is a fact now generally recognized, that formations of analogous composition were produced in succession at periods very distant from each other. The doubt which I mention without participating in it bears on a point much less certain; whether the mica schists, indubitably placed in the midst of primitive rocks (under those in which organic remains begin to appear), are newer than the secondary rocks of another country. I confess that in the part of the globe which I have been able to examine, I have seen nothing that could warrant this opinion. Granular syenitic rocks, repeated twice, perhaps even three times, in primitive, intermediary, (and secondary?) formations, are analogous phenomena that are become familiar to us within the last fifteen years; but the want of agreement in the age of the great homonymous formations appears to me far from being sufficiently proved by direct observations on the contact of superposed beds. The chalk or the Jura limestone may cover immediately the primitive granite on one side, and be separated on the other by numerous secondary and transition rocks; these very common facts prove only the subtraction, the absence, the non-developement of several intermediary members of the geognostical series. Grauwacke may, on one hand, dip under a feldspathic rock, for instance, beneath a transition granite, or the zircon syenite; and, on the other hand, it may be superposed on a black limestone with madrepores. This situation denotes only the included position of a bed of grauwacke between limestone rocks and feldspar transition rocks. Since, in consequence of the important researches of MM. Cuvier and Brongniart, a profound examination of fossil organic bodies has diffused new spirit into the study of tertiary deposits, the discovery of the same fossils in the analogous beds of very distant countries has rendered still more probable the *isochronism* of widely extended formations.

It is by this isochronism only, this admirable order of succession, we are enabled to observe with certainty. The attempts which have been made by the Hebraic geologists to subject the epochas to absolute measures of time, and to connect the chronology of antient cosmogonic traditions with actual observations of nature, have proved fruitless. "It has more than once been desired," says M. Ramond, in a discourse abounding with philosophical views, "that we could find a supplement to our short annals, in the monuments of nature. The historical ages might, however, have sufficed to teach us, that the succession of physical and moral events is not regulated by the uniform progress of time, and cannot in consequence furnish its measure. We see, in looking back, a succession of creations and destructions, by the various arrangements of the beds that form the crust of the globe. They give us the idea of several distinct epochas; but these epochas, so fertile in events, may have been very short compared to the number and the importance of the results. Between the creations and destructions, on the contrary, we perceive nothing, whatever might be the immensity of the intervals; there every thing is lost in the mist of an undeterminable antiquity, the degrees of which cannot be appreciated, because the succession of phenomena has no scale that can be referred to the division of time." (*Mémoires de l'Institut*, 1815, p. 47.)

In the geognostical monography of a district of small extent, the environs of a town for instance, we cannot mark too minutely the different beds that compose the local formations. Beds of sand and clay, the subdivisions of gypsums, strata of marly and oolitic limestone known in England by the names of Purbeck-beds, Portland-stone, Coral-rag, Kelloway-rock, and Cornbrash, then acquire great importance. Thin beds of secondary or tertiary formations, containing assemblages of fossil bodies well characterized, have served the purpose of an *horizon* to the geognost. He has been enabled, as the beds are prolonged, to connect with one of them what is placed above or below in the order of the whole series. Even the particular denominations by which these beds are distinguished are of great advantage in a geognostic description, however strange or improper may be their signification or their origin as drawn from the language of miners. But when we treat of the place of rocks on a surface of great extent, it becomes indispensable to consider the formations or usual groupings of certain beds under a more general point of view. We must then be more prudent and circumspect in the distinction and nomenclature of rocks. The work of M. Freiesleben on the plains of Saxony, which are more than seven hundred square leagues in extent (*Geogr. Beschr. des Kupferschiefergebirges, in* 4 *Th.* 1807—1815.) exhibits an admirable model of the union of local observations with geognostical generalizations. These generalizations, and attempts to simplify the table of formations, and to dwell only on the great characteristic features, should be made with more or less timidity, according as the objects described are the basin of a river, an insulated province, a country as large as France or Germany, or an entire continent.

The more profoundly we study the nature of rocks, the more we perceive the connection between formations which at first appeared altogether independent to be made evident by the great phenomena of *alternation*, that is, by a periodical succession of beds that have a certain analogy in their composition, and sometimes also in the contained fossil bodies. Thus in the transition mountains, for instance in America, (at the entrance of the plains of Calabozo,) the beds of greenstone and euphotide; in Saxony (near Friedrichswalde and Maxen), the schists with ampelite, grauwackes, porphyries, the black limestones and greenstones, constitute, after frequent and repeated *alternations*, the same formation. It often happens that the subordinate beds appear only at the extreme limit of a formation, and then assume the aspect of an independent rock. The copper and bituminous marles (Kupferschiefer) which are found in Thuringia between the alpine limestone (zechstein) and the red sandstone (rothe liegende), in which mines have been worked for ages, are *represented* in several parts of Mexico, New Andalusia, and in the south of Bavaria, by numerous beds of marly clay, more or less carburetted and imbedded in alpine limestone. Similar circumstances often give to gypsums, to sandstones, and to small beds of compact limestone, the appearance of particular formations. We perceive their dependance or their *subordination* by their frequent association with other rocks, their want of extent and thickness, or by their total suppression, which has been frequently observed. We must not forget (and I have been struck with this fact in both hemispheres,) that the great calcareous formations, for instance, the alpine limestone, have *their sandstones*, as the sandstones very extensively distributed have *their limestone beds*. Thin layers of sandstone, of limestone, and of gypsum, characterize, in every zone, the deposits of coal and of rock-salt or muriatiferous clay (*salzthon*): these insulated deposits are most frequently covered only by small local formations. By neglecting these considerations, which must be familiar to every experienced geognost, the type of the great independent formations has been rendered too complicated.

The phenomenon of *alternation* is manifested, either locally in rocks superposed several times on one another and constituting one complex formation, or in the series of formations considered in a general view. Either greenstones, or syenites, schists and transition limestones, beds of limestone and marl, alternate immediately; or a whole system of mica-slates and feldspar granular rocks (granites, gneiss, and syenites), appears again amidst the transition formations, and separates the grauwackes and the limestones with orthoceratites, from the primitive homonymous system. The first knowledge of this fact, one of the most important and most unexpected in modern geognosy, we owe to the excellent observations of MM. Léopold de Buch, Brochant, and Haussmann. This phenomenon connects in some degree the transition with the primitive formations, not with respect to time or relative antiquity, but with regard to analogy of composition and aspect. The fact, that very ancient granular rocks, entirely destitute of organic remains, succeed to compact rocks containing organic remains, has led some distinguished geognosts to conclude, that this *alternation* of rocks containing shells with others free from shells might perhaps extend beyond what we call primitive formations. They have not merely enquired if clay-slate, mica-slate, and gneiss, did not support granites which have been considered as the most ancient rocks: but the question has also been agitated, whether grauwacke, and black limestones with madrepores, might not be found beneath granites? According to this view, the primitive and transition-series would form but one class of rocks; and the former might be regarded as being interposed in a formation that was posterior to the developement of organized beings, and which penetrates to an unknown depth in the interior of the globe. I believe that no direct observation can as yet be adduced in favour of these suppositions. The fragments of rocks which I saw imbedded in the lithoïd lavas of the volcanoes of Mexico, Quito, and Vesuvius, and that are thought to have been torn from the bowels of the earth, seem to belong to altered rocks of granite, mica-slate, syenite, and granular limestone, and not to grauwacke and limestones with madrepores.

In the tabular arrangement of rocks, the great divisions, known by the name of primitive, intermediary, secondary, and tertiary formations, are preserved. The natural limits of these four *systems of rocks* are, the clay-slate with ampelite and lydian-stone, alternating with compact limestones and grauwacke, the coal formation, and those that succeed immediately to chalk. In geognosy, as in descriptive botany (phytography), the subdivisions or small groups of families have more distinct characters than the great divisions or classes. This happens in every science in which we ascend from the individual to the species,

the primitive and transition rocks, we shall denote by the name of volcanic formations, the *least interrupted series* of rocks altered by fire.

In making the enumeration of rocks, I have used the names most generally employed by the geognosts of France, Germany, England, and Italy. I apprehended, that in endeavouring to perfect the nomenclature of formations, I might add new difficulties to those which already exist in discussing the subject of superpositions. I have, however, carefully avoided the denominations which have been too long preserved, of *lower and upper limestone; gypsum of the first, second, or third formation; old or new red sandstone*. These denominations no doubt present a true geognostic character; they relate, not to the composition of rocks, but to their relative age. As the general type, however, of formations in Europe, cannot be modelled on that of a single canton, the necessity of admitting parallel formations (*sich vertretende Gebirgsarten*) renders the names of *first* or *second gypsum*, of *old* or *middle sandstone*, extremely vague and obscure. In one country we are justified in considering a bed of gypsum or sandstone as a particular formation, while in another it must be regarded as subordinate to neighbouring formations. *Geographical denominations* are certainly the best, and they give very precise ideas of superposition. When we say that a formation is identical with the porphyry of Christiania, the lias of Dorsetshire, the sandstone of Nebra (*bunte sandstein*), the coarse marine limestone (calcaire grossier) of Paris, those assertions leave no doubt in the mind of a well-informed geognost of the position we mean to assign to the formation described. By a tacit convention, therefore, the words, zechstein of Thuringia, limestone of Derbyshire, calcaire grossier of Paris, have been introduced into mineralogical language; they call to mind a limestone which immediately succeeds to the old red and coal sandstone, a transition limestone placed beneath the coal sandstone, or, formations more recent than chalk. The only difficulty in this multiplicity of geographical denominations consists in the choice of names, and in the degree of certainty which we have acquired respecting the place, or relative age of the rock, to which we refer the others. The English geognosts seek their *red marl* and their *lias* on the Continent; the German geognosts their *variegated sandstone* and their *muschelkalk*. These terms are associated in the minds of travellers with local recollections; consequently, in order to create precise ideas, we have only to chuse localities that are generally known and celebrated either by the working of mines, or by scientific descriptions.

In order to diminish the effect of national vanity, and annex the new names to more important objects, I long since proposed (1795) the denomination of Alpine and Jura limestone. A part of the High-Alps of Switzerland and the greater part of the Jura are no doubt formed of those two rocks; the names, however, now generally received, of Alpine limestone (*zechstein*) and Jura limestone, ought in my opinion to be modified or altogether abandoned. The lower beds of the Jura mountains, containing gryphites, belong to a more ancient formation; and a great part of the Alps of Switzerland is certainly not zechstein, but transition limestone, according to MM. de Buch and Escher. It would be, therefore, better to choose the geographical names of rocks, from those of insulated mountains, the whole visible mass of which belong to one general formation, than to borrow them, as I have done, improperly from the whole chain. I believed, and many geognosts were of this opinion, that the Jura limestone (limestone of the caverns of Franconia) was generally, on the Continent, placed below the sandstone of Nebra (*bunte sandstein*), and between this sandstone and the zechstein. Subsequent observations have proved, that the name of *Jura limestone* had been properly applied to rocks which are distinctly separated from the mountains of Western Switzerland; but the principal geognostical place of this formation (when the lower formations are not wanting) is far above the sandstone of Nebra, between muschelkalk (or quadersandstein) and the chalk. A geographical name, when applied with propriety to several analogous rocks, calls our attention to the identity of their geognostic positions; but the place which those homonymous rocks should occupy in the whole series is never well determined, but when the geographical name has been fixed upon after a complete certainty with respect to their geognostic position. Geognosts find themselves in similar circumstances, when endeavouring to determine the relative age of the molasse of Argovia (nagelfluhe), and the quadersandstein of Pirna (white sandstone of M. Bonnard); two very recent rocks, which have been studied well separately, but the connection of which with each other, with the chalk, and with the Jura limestone have but lately been made out. We may feel sufficiently confident in having ascertained that rocks in the New Continent are identical with the molasse or the quadersandstein, although we cannot determine their relations with the rest of the secondary, or tertiary beds. When formations are not in immediate contact, and are not covered by *beds* of which the position is known, we can only judge of their relative age from simple analogies.

The *terms* of the geognostic series are *simple* or *complex*. The greatest part of the primitive formations, as granite, gneiss, mica-slate, clay-slate, &c. belong to the simple terms. The complex terms are found chiefly among transition rocks, in which each formation comprehends a complete group of rocks that alternate periodically. The terms of the series are not transition limestone, nor grauwacke, constituting independent formations; but associations of clay-slate, greenstone, and grauwacke; of porphyry and grauwacke; of steatitic granular limestone, and pudding-stones formed from primitive rocks; and of clay-slate and black limestone. When those associations are composed of three or four alternating rocks, it is difficult to give them names sufficiently expressive to indicate the whole composition of the group, and all the particular members of the complex term of the series. But it may help to fix the groups in the memory, or to call to mind the rocks that predominate, and which are never quite wanting in the neighbouring groups. Thus, steatitic granular limestone characterises the formation of the Tarentaise; grauwacke, the great transition formation of the Hartz and the banks of the Rhine; metalliferous porphyries, containing much hornblende, and almost destitute of quartz, the formation of Mexico and Hungary. Though the phenomena of alternation and of grouping attain their maximum in the transition formations, yet they are not entirely excluded from primitive and secondary formations; in both, the complex terms are blended with the simple terms of the geognostic series. I shall mention among the secondary formations, the sandstone placed above the alpine limestone (sandstone of Nebra, *bunte* sandstein), which is an association of marly clay, sandstone, and oolites; the limestone that covers red sandstone with coal (the zechstein, or alpenkalkstein), which is an association less constant of limestone, of muriatiferous gypsum, stinkstein, and friable bituminous marl (asche of the Mansfeld miners). In primitive formations we find the first three terms of the series, the most ancient rocks, either insulated, or alternating two and two, according as they are geognostically nearer each other by their relative age or all three alternating together. Granite sometimes forms constant associations with gneiss, and gneiss with mica-slate. These alternations follow particular laws; we see, for instance, in Brazil, and (although less distinctly) in the chain on the coast of Venezuela, granite, gneiss, and mica-slate in triple association. But I have no knowledge of granite alternating only with mica-slate, or of gneiss and mica-slate alternating only with clay-slate.

We must not confound (a point on which I have often insisted in this essay) rocks passing insensibly to those with which they are in immediate contact; for instance, mica-slates that *oscillate* between gneiss and clay-slate, with rocks that alternate together, and preserve all their distinct characters of composition and structure. M. d'Aubuisson has long since shown, that chemical analysis connects clay-slate with mica. (*Journal de Physique*, tom. lxviii. page 128. *Traité de Géognosie*, tom. ii. page 97.) The former has not, indeed, the metallic lustre of mica-slate: it contains a little less potash and more carbon; the silica does not unite in knots or thin plates of quartz, as in mica-slate; but there can be no doubt that the scales of mica constitute the principal basis of clay-slate. These scales are so united together, that the eye cannot distinguish them in the mass. Perhaps this very affinity prevents the alternation of clay-slate and mica-slate; for in these alternations nature seems to favour the association of heterogeneous rocks, or, to use a figurative expression, she delights in those associations in which alternating rocks exhibit a great contrast of crystallization, mixture, and colour. I saw in New Spain dark-coloured greenstone alternating thousands of times with reddish-white syenites, that abound more in quartz than in feldspar; veins of syenite occur in this greenstone, and veins of greenstone in the syenite; but neither of those rocks passes into the other. (*Essai Politique sur la Nouvelle Espagne*, tom. ii. p. 523.) At their actual contact, they exhibit differences as strongly marked as the porphyries that alternate with grauwackes and syenites, the black limestone that alternates with transition clay-slate, and many other rocks altogether heterogeneous in their composition and general aspect. It may be observed further, that when in primitive formation, rocks more nearly allied to each other from their composition than by their structure, or their mode of aggregation, for instance, granite and gneiss, or gneiss and mica-slate, alternate, those rocks do not show the same tendency to pass into each other, which they display in non-complex formations. We have already observed above, that a bed β, when it occurs more frequently in the rock α, points out to the geognostic traveller, that the simple formation α is about to be succeeded by a complex formation, in which α and β alternate. Afterwards it may happen that β assumes a greater development; that α is no longer an alternating rock, but merely a subordinate bed to β, and that this rock β appears by itself, till, by the frequent appearance of beds γ, it forms the prelude to a complex formation of β alternating with γ. We may substitute to these signs the words granite, gneiss, and mica-slate; those of porphyry, grauwacke, and syenite; of gypsum, marl, and fetid limestone (*stinkstein*); but the *pasigraphic* language has the advantage of generalising problems, and conforms more to the requisites of *geognostic philosophy*, of which I here attempt to trace the first elements, as far as regards the study of the superposition of rocks. But, between formations, simple and nearly connected in the order of their relative antiquity, between the formations α β γ, complex formations, α β and β γ, are often found placed, (that is, α alternating with β, and β alternating with γ); it is also observed, though less frequently, that one of those formations (for instance, α) takes an increase so extraordinary, that it envelopes the formation β, and that β, instead of appearing as an independent rock, placed between α and γ, is only a bed in α. Thus, in Lower Silesia, the red sandstone contains the formation of zechstein; the limestone of Kunzendorf with impressions of fish, and analogous to the bituminous marl of Thuringia abounding also in fish is entirely enveloped in the coal sandstone. (Buch, *Beob.* t. i. p. 104. 157.; Id. *Reisenach Norwegen*, t. i. p. 158.; Raumer, *Gebirge von Nieder-Schlesien*, p. 79.) M. Beudant (See *Min.* t. iii. p. 183.) has observed a similar phenomenon in Hungary. In other regions, for instance, in Switzerland, and at the southern extremity of Saxony, the red sandstone disappears altogether; being replaced, or in some sort suppressed, by an immense development of grauwacke, or of alpine limestone. (Freiesleben, *Kupfersch*, b. iv. 109.) These effects of the alternation and unequal development of rocks are so much the more worthy of attention, as the study of them may throw light on some apparent deviations from a type of superposition generally received, and may serve to refer to a common type the series of positions observed in countries very distant from each other.

To designate the formations composed of two rocks that alternate with one another, I have generally preferred the words *granite* and *gneiss*, *syenite* and *greenstone*, to the more usual expressions of *granite-gneiss*, *syenite-greenstone*; I apprehended that the latter method of designating the formations composed of alternating rocks, might rather suggest the idea of a passage of granite to gneiss, or of syenite to greenstone. In fact, a geognost whose researches on the trachytes of Germany have not been sufficiently appreciated, Mr. Nose, has already employed the words *granite-porphyries*, and *porphyry-granites*, to indicate the varieties of structure and aspect, and to separate the porphyroid granites from porphyries, which, by the frequency of crystals imbedded in the mass, exhibit a structure of aggregation, a real granitic structure. In adopting the denominations of granite and gneiss, of syenite and porphyry, of grauwacke and porphyry, of limestone and clay-slate, no doubt is left on the nature of the complex terms of the geognostic series.

Among the different proofs of the identity of formations in the most distant regions of the globe, one of the most striking, and which we owe to zoology, is the identity of the organic bodies buried in beds of similar position. The researches that lead to this species of proof have greatly exercised the sagacity of the learned, since MM. de Lamarck and Defrance began to determine the fossil shells in the vicinity of Paris, and since MM. Cuvier and Brongniart have published their well known work on fossil bones and tertiary beds. As the most considerable mass of the formations that compose the crust of our planet contains no vestiges of organised bodies, and as those vestiges are very rare in transition rocks, are often broken, and difficult to separate from the rock, in the oldest secondary beds, the profound study of fossil bodies comprehends but a small part of geognosy, but a part which is highly deserving of the attention of the philosopher. The problems to be solved are numerous; they relate to the geography of animals, the races of which are extinct, but which still belong to the history of our planet; and they involve the description of the zoological characters, by means of which we wish to distinguish the different formations. In order to adhere to my plan of considering, in this *Introduction to the tabular Arrangement of Rocks*, the different objects in the most general point of view, I shall mention the queries in geognostic zoology, that appear the most important in the present state of the science, and of which the solution has been attempted with more or less success. What are the genera, and (if the state of preservation and the slight adherence to the rocky mass permit a more complete determination,) what are the species to which fossil remains belong? Will not an exact determination of the species lead us to recognise with certainty those which are identical with the plants and animals of the present world? What are the classes, the orders, and the families of organised beings, which exhibit the greatest number of those analogies? In what proportion does the number of identical genera and species augment, as the rocks or earthy deposits are newer? Is the order observed in the superposition of intermediary, secondary, tertiary, and alluvial formations, every where in harmony with the increasing analogy which the types of organisation exhibit? Do these types succeed each other from below upwards, (passing from grauwacke and black transition limestone, by coal sandstone, alpine and Jura limestone, and chalk, to tertiary gypsum, freshwater formations, and modern alluvia,) in the same order which we adopt in our systems of natural history, arranging the several beings as their structure becomes more complicated, and as other systems of organs are found added to the organs of nutrition? Does the distribution of organised fossil bodies indicate a progressive development of vegetable and animal life on the globe; a successive appearance of acotyledon and monocotyledon plants, of zoophites, crustacea, molluscæ, (cephalopodes, acephali, gasteropodes,) of fish, saurians (oviparous quadrupeds), dicotyledon plants, and sea and land mammiferæ? In considering fossil bodies, not in their connection with certain rocks in which they have been discovered, but merely in relation to their distribution according to climates, is an appreciable difference found between the species which predominate in the ancient and in the new continent, in temperate climates and in the torrid zone, in the northern and southern hemispheres? Is there every where found a certain number of the tropical species, which seem to denote that, independently of a distribution of climates similar to those of the present time, they have experienced in the first period of the world, the high temperature which the fissured crust of the globe, strongly heated in its interior, had given to the circumambient air? Is it possible to distinguish, by precise characters, fresh-water from marine shells? Is the determination of the genera complete, or are there (as among fish,) some genera of which the species live alike in rivers and the seas? Although in some tertiary rocks fluviatile shells are found mixed, as at the mouths of rivers, with pelagic shells, may it not be observed, in general, that the first form particular deposits, characterising those formations, the study of which had been hitherto neglected, and which are of a very recent origin? Have any fresh-water shells ever been discovered beneath Jura limestone, near fish reputed to be fluviatile, in the bituminous slate of the alpine limestone? Are the same species of fossils found in the same formations on different points of the globe? Can they furnish zoological characters by which the various superimposed formations may be recognised? or ought we not rather to admit that those species which the zoologist must regard as identical, according to the adopted methods, penetrate through several formations, and appear even in those which are not in immediate contact? Ought not the zoological characters to be drawn from the total absence of some species, and from their relative frequency or *predominance*; finally, from their constant association with a certain number of other species? Ought we to divide a formation, the unity of which has been recognised, from the relative position and the identity of the beds which are interposed equally in the upper and lower strata, for the sole reason that the former strata contain fresh-water shells, and the latter sea shells? Is the total absence of organised bodies in certain masses of secondary and tertiary formations, a sufficient reason for considering those masses as particular formations, if other geognostic facts do not justify that separation?

A part of these problems has been presented long since to naturalists. Lister maintained above a hundred and fifty years ago, that every rock was characterised by different fossil shells. (*Phil. Trans.*, No. 76., p. 2283.) To prove that the shells of our seas and lakes are specifically different from fossil-shells (*lapides sui generis*), he affirms that the latter, those, for instance, of the quarries of Northamptonshire, bear all the characters of our *Murex*, our *Tellina*, and our *Trochus*; but that those naturalists, who are not contented with a vague and general view of things, will find that many fossil shells are specifically different from all the shells of the present world." Nearly at the same period, Nicholas Stenon (*de solido intra solidum contento*, 1669, p. 2. 17. 28. 63. 69. fig. xx.—xxv.) first distinguished " the primitive rocks anterior to the existence of plants and animals on the globe, and consequently never containing any organic remains, from the secondary rocks superimposed on the former, and filled with animal remains (*turbidi maris sedimenta sibi invicem imposita*)." He considered every bed of secondary rock " as a sediment deposited by an aqueous fluid;" and adopting a theory quite similar to that of Deluc, " on the formation of vallies by longitudinal sinkings, and on the inclination of beds originally horizontal;" he admits for the formations of Tuscany, in the same manner as modern geologists, " six great epochas of nature (*sex distinctæ Etruriæ facies, ex præsenti facie Etruriæ collectæ*), according as the sea periodically inundated the continent, or retired within its ancient limits." At the time when the observation of nature gave rise in Italy to the first ideas on the relative age, and the succession, of primitive and secondary beds, zoology and geognosy could not as yet lend mutual aid to each other, because the zoologists had no knowledge of rocks, and the geognosts were altogether strangers to the natural history of animals. They vaguely regarded every thing as specifically identical that exhibited any analogy of form; but, at the same time, (and that was a step made in the right path,) they paid attention to the fossils which abounded in particular rocks. Thus the denominations of gryphite limestone, limestone with trochites, schists with ferns, schists with trilobites (Gryphiten-und Trochiten-Kalk; Kräuterund Trilobiten-Schiefer,) were very anciently employed by the mineralogists of Germany. The determination of the genera characterised by the teeth, the grooves, the sharp and indented edges of the hinge, by the folds and fillets at the opening of the shell, is much more difficult in the most ancient secondary rocks than in the tertiary formations; the former being generally less friable, and adhering more to the shell of the fossil body. This difficulty augments when we wish to distinguish the species, and becomes almost insurmountable in some calcareous transition rocks, and in the muschelkalk which contains broken shells. If the zoologic characters of a certain number of formations could be formed from genera sufficiently distinct, if trilobites and orthoceratites belonged exclusively to intermediary formations, the gryphites to the alpine limestone (zechstein), the pectinites to the bunte-sandstein (sandstone of Nebra), trochites and mytilites to the muschelkalk, tellinæ to quadersandstein, ammonites and turritellæ to the limestone and marl of the Jura, the ananchytes and the spatangi to the chalk, and cerithia to the calcaire grossier of Paris; the knowledge of those genera would greatly facilitate the determination of rocks; it would no longer be necessary to examine the superposition of formations on the spot; this might be ascertained in the cabinet, and only by consulting collections. But Nature has not rendered so easy the study of the shelly masses that constitute the crust of our planet. The same types of organisation are repeated at very different epochas; the same genera are found in the most distinct formations. Orthoceratites occur in transition limestone, alpine limestone, and variegated sandstone; terebratulites in the limestone of Jura and the muschelkalk; trilobites in transition clay-slate, in the bituminous slate of the zechstein, and, according to an excellent geognost, M. de Schlottheim, even in Jura limestone; there are pentacrinites in transition clay-slate, and the most modern muschelkalk. The ammonites penetrate through many calcareous and marly formations, from grauwacke, (Raumer *Versuche*, p. 22.; Schlottheim, *Petrefactenbaude*, p. 38.) as far as into the lower beds of the chalk. Trunks of monocotyledon plants occur in the red sandstone, and in the marls of the fresh-water gypsum formed at a period when dicotyledon plants were already abundant in the world.

But in the present age, naturalists are no longer satisfied with vague and uncertain notions, and they have sagaciously observed that the greatest number of those fossils (gryphites, terebratulites, ammonites, trilobites, &c.) buried in different formations, are not specifically the same; that many species which they have been enabled to examine with precision, vary with the superposed rocks. The fish that have been observed in transition slate (Glaris), in the bituminous slate of the zechstein, in the Jura limestone, in the tertiary limestone with cerithia of Paris and of Monte Bolca, and in the gypsum of Montmartre, are distinct species, partly pelagic, and partly fluviatile. Ought we to conclude from this assemblage of facts, that all the formations are characterised by particular species? that the fossil shells of the chalk, of the muschelkalk, of the Jura limestone, and of the alpine limestone, all differ from each other? This would be, in my opinion, to carry the induction much too far; and M. Brongniart himself, who knows so well the value of zoological characters, restricts their absolute application to the case where it is not opposed by superposition, or the circumstances of their geognostic place. I might mention the cerithia of the limestone which is found (near Caen) below the chalk, and which seem to indicate (like the repetition of clays with lignites above and below the chalk,) a certain connection between formations which at first sight appear to be quite distinct. I might dwell on other species of shells which belong at the same time to several tertiary formations, and add, that even if at a future time we should succeed in separating, by slight characters, and insensible gradations, species which are at present considered as identical, those nice distinctions would not quite satisfy the enquirer concerning the universality so desirable, of zoologic characters in geognosy. Another objection, drawn from the influence which climates exert even on pelagic animals, appears to me still more important. Although the seas, from well-known physical causes, have, at immense depths, the same temperature at the equator and within the temperate zone, yet we see, in the present state of our planet, the shells of the tropics (among which the univalves predominate, as they do among the testaceous fossils,) differ much from the shells of northern climates. The greatest number of those animals adhere to reefs and shallows; whence it follows, that the specific differences are often very sensible in the same parallel on opposite coasts. Now, if the same formations are repeated and extended to immense distances, from east to west, from north to south, and from one hemisphere to the other, is it not probable, whatever may have been the complicated causes of the antient temperature of our globe, that variations of climate must have modified, heretofore as now, the types of organisation; and that the same formation (that is, the same rock placed in the two hemispheres between two homonymous formations,) would have enveloped different species? It no doubt often happens, that superposed beds present a striking difference in their fossil organic remains. But can we thence conclude, that after a deposit was formed, the beings which then inhabited the surface of the globe were all destroyed? It is incontestable that generations of different types have succeeded to one another. The ammonites, which are scarcely to be found among transition rocks, attain their *maximum* in the beds that represent, on different points of the globe, the muschelkalk and Jura limestone; they disappear in the upper beds of the chalk, and above that formation. The echinites, extremely rare in alpine limestone, and even in muschelkalk, become on the contrary very common in the Jura limestone, chalk, and tertiary formations. But nothing proves that this succession of different organic types, this gradual destruction of genera and species, coincides necessarily with the periods at which each formation took place. "The consideration of the similitude or of the difference between organic remains is not of great importance, (says M. Beudant, *Voyage Min.*, tom. iii. p. 278.) when we compare the deposits which have been formed in countries very distant from each other; but is highly important if we compare deposits that are near together."

In opposing the absolute opinion which we might be tempted to form on the value of zoological characters, I am far from denying the important services rendered to geognosy by the study of fossil organic bodies, if we consider that science in a philosophic point of view. Geognosy is not confined to the research for diagnostic characters; it comprehends the whole of those relations in which we may consider every formation: 1st, its position; 2dly, its oryctognostic constitution (that is, its chemical composition, and the particular mode of aggregation, more or less crystalline, of its molecules); 3dly, the association of different organised bodies that are found imbedded in it. If the superposition of different heterogeneous rocky masses exhibits to us the successive order of their formation, why should we not be interested in knowing the state of organic nature at the different epochas when those deposits were formed? There can be no doubt that on a surface of several thousand square leagues (in Thuringia, and in all the northern part of Germany), nine superposed formations, viz. that of transition limestone, grauwacke, red sandstone, zechstein with bituminous slate, muriatiferous gypsum, oolitic limestone, gypsum in clay, muschelkalk, white sandstone (quadersandstein), have been recognised as distinct, without having had any recourse to the use of zoological characters; but it does not follow from thence, that the most minute examination of those characters, or rather that the most intimate knowledge of the fossils contained in each formation, is not indispensable in order to form a complete geognostic work. The study of formations is similar to that of organised beings. Botany and zoology, considered at present in a more elevated point of view, are no longer confined to the examination of some external characters, and distinctions of species; those sciences enter more profoundly into the study of the whole vegetable and animal organisation. The characters drawn from the forms of shells suffice to distinguish the different species of testaceous acephalæ. Shall we, on that account, regard as superfluous the knowledge of the animals which inhabit these shells? Such is the connection among phenomena, and their natural relations (those of life as well as those displayed by the stony deposits formed at different periods), that if we neglect any of them we shall form not only an incomplete, but most frequently, an erroneous picture of the whole.

In the case of the conformity of position, there may be identity of mass (that is, of mineralogical composition,) and diversity of the fossils, or diversity of mass and identity of fossils. The rocks β and β, placed at great horizontal distances, between two identical formations, α and γ, belong either to the same formation, or are parallel formations. In the first case, their composition is similar; but on account of the distance of places, and effects of climate, the organic remains which they contain may differ considerably. In the second case the mineralogical composition is different, but the organic remains may be analogous. I consider the words *identical formations, parallel formations*, as indicating the conformity or non-conformity of mineralogical composition; but they do not enable us to prejudge respecting the identity of fossils. If it be sufficiently probable that the deposits β and β, placed at great horizontal distances between the same rocks, α and γ, are formed at the same epochs, because they contain the same fossils, and are of the same composition, it is not equally probable that the *epochas of formation* are very distant from each other, when the fossils are distinct. It may be conceived that in the same zone, in a country of small extent, generations of animals have succeeded each other, and have characterised, as it were by particular types, the *epochs* of formations; but at great horizontal distances, beings of various forms may have occupied simultaneously, in different climates, the surface of the globe, or the basin of the sea. It may be observed further, that the position of β between α and γ proves that the formation of β is anterior to that of γ, and posterior to that of α; but nothing gives us an absolute measure of the intervals of time; and different insulated deposits of β may not be simultaneous.

It seems to result from the facts collected through the zeal and sagacity of naturalists within a few years, that, though we must not always expect to find, as Lister pretended, in every different formation, vestiges of different organised bodies, yet those formations which are considered as identical from their place and composition, usually contain, in the most distant countries of the globe, associations of species entirely similar. M. Brongniart (whose labours, joined to those of MM. Lamarck, Defrance, Beudant, Desmarest, Prevost, Ferussac, Schlottheim, Wahlenberg, Buckland, Webster, Phillips, Greenough, Warburton, Sowerby, Brocchi, Soldani, Cortesi, and other celebrated mineralogists, have so much advanced the study of *subterraneous conchology*,) has recently pointed out the striking analogies which fossil bodies present in certain formations of Europe and North America. He has attempted to prove that a formation is sometimes so disguised, that it can only be recognised by zoological characters. (Brongniart, *Hist. Nat. des Crustacées Fossiles*, p. 57. 62.) In the study of formations, as in all the descriptive physical sciences, it is the sum of many characters that should guide us in the search after truth. The specific description of the vestiges of plants and animals contained in the various rocks becomes a sort of *Flora* or *Fauna*. Now, in the primordial world, as in the present, the vegetation and the animal productions of various portions of the globe appear to be less characterised by some insulated forms of an extraordinary aspect, than by the association of many forms specifically different, but analogous among themselves, notwithstanding the distance of the localities. In discovering a new land, near to the streight of Torres, it would not be easy to determine, from a small number of productions, if that land be contiguous to New Holland, to one of the Molucca islands, or to New Guinea. To compare formations with relation to fossils, is to compare the *Floras* and *Faunas* of various countries at various periods; it is to solve a problem so much the more complicated, as it is modified at once by space and time.

Among the zoological characters applied to geognosy, the absence of some fossils often characterises formations better than their presence. This is the case with the transition rocks; we find, in general, only madrepores, encrinites, trilobites, orthoceratites, and shells of the family of terebratulæ; that is to say, fossils, of which some species not identical, though analogous, are found in very modern secondary beds; but in these transition rocks many other remains of organised bodies are wanting, which appear in abundance above the red sandstone. The opinion which we may form from the absence of certain species, or on the total absence of organic bodies, may, however, be the consequence of an error which it will be useful to point out. In examining in a general manner the formations containing shells, we see that organised bodies are not always equally distributed in the mass: but 1st, that strata entirely destitute of fossils alternate with other strata where they abound; 2dly, that in the same formation, particular associations of fossils characterise certain strata that alternate with other strata containing different fossils. This phenomenon, long since noticed, is found to occur in the muschelkalk, and alpine limestone (zechstein), which are often separated from the coal sandstone by a bed containing trochites. (Buch, *Beob.*, t. i. p. 155, 156. 171.) This is also the case in the Jura limestone, and in several tertiary formations. In studying only the chalk in the vicinity of Paris, we might almost be induced to think that univalve shells are entirely wanting in this formation; but the polythalamous univalves, the ammonites, as we have already observed, are very common in England, in the oldest beds of the chalk. Even, in France, (on the hill of Saint Catherine, near Caen,) the tufaceous chalk (craie tuffeau) and the chalk with chlorite (craie chloritée) contain many fossils that are not found in the white chalk. (Brongniart, *Caractères Zool.*, p. 12.) The formations not being equally developed in different countries, portions of formations being often taken for entire and complete formations, those which are destitute of shells in one region, may contain them in another. This consideration is important in order to prevent the disposition to multiply formations unnecessarily; for when on the same point of the globe, a formation (for instance, sandstone,) contains in its lower part a great quantity of fossils, which are entirely wanting in the upper part, this absence of fossils does not alone justify the division of the formations into two distinct portions. M. Brongniart, in the geological description of the environs of Paris, has properly united the meuliere without shells with those beds which are completely filled with fresh water shells.

We have just seen that a formation may contain, in different strata, petrifactions of different species, but that most frequently some species of an inferior stratum will be found in the mass of heterogeneous species that occurs in the superimposed bed. When this difference bears upon genera, some of which are pelagic shells, and others fresh-water shells, the problem of the unity, or the indivisibility of a formation becomes more embarrassing. We must first distinguish two cases, that in which some fluviatile shells are found mingled with a great number of marine shells, and that where marine and fluviatile shells may alternate bed by bed. MM. Gillet, De Laumont, and Beudant have made interesting observations on the mixture of marine and fresh-water productions in the same bed. M. Beudant has proved by ingenious experiments, that many fluviatile molluscæ may be gradually accustomed to to live in water which has all the saltness of the ocean. The same naturalist has examined, conjointly with M. Marcel de Serres, certain species of paludinæ, which, preferring brackish water, are found near our coasts, sometimes with pelagic shells, sometimes with fluviatile shells. (*Journal de Phys.*, t. lxxxiii. p. 137., t. lxxxviii. p. 211.; Brongniart, *Geog. Min.*, p. 27. 54. 89.) To these curious facts others may be added, which I have published in the Narrative of my *Voyage aux Regions Equinoxiales*, (t. i. p. 556., t. ii. p. 606.), and which seem to explain what formerly took place on the globe, from what we now observe. I saw crocodiles on the coast of Terra Firma, between Cumana and Nuova Barcelona, advance far into the sea. Pigafetta made the same observation on the crocodiles of Borneo. To the south of the Isle of Cuba, in the gulph of Xagua, there are lamantins in the sea, where springs of fresh water issue in the midst of the salt water. When we reflect on the whole of these facts, we are less astonished at the mixture of some land productions with many others that are incontestably marine.

The second case which we have mentioned, that of alternation, never occurs I believe in so decided a manner, as the alternation of clay-slate with black limestone in the same transition formation, or, (to recall to mind a fact, which relates to the distribution of organised bodies,) than the alternation of two great marine formations (limestone with cerithia and the sandstone of Romainville), with two great fresh-water formations (the gypsum, and meuliere of the plateau of Montmorency). What the accurate observation of superpositions has hitherto supplied may be reduced to alternating beds of gypsum and marl, placed between two marine formations; and containing in the middle, (where they are thickest,) land and fresh-water productions; and towards the upper and lower limits, marine productions, both in gypsum and in marl; such is the geological constitution of the gypsum of Montmartre. The specific variations in the petrifactions, the mixture observed at Pierrelaie, and the phenomenon of alternation that occurs at Montmartre, are not sufficient to authorise the subdividing of a formation. The marls and the gypsum which contain marine shells, (No. 16 of the third mass,) cannot be geognostically separated from the marl and gypsum containing fresh-water productions. MM. Cuvier and Brongniart, therefore, have not hesitated to consider the whole of those marls, with the marine and fresh-water gypsum, as the same formation. Those naturalists have even cited this assemblage of alternating beds as one of the clearest examples of what should be understood by the word *formation*. (*Geogr. Miner*, p. 31. 39. 189.) In fact, different systems of beds may be contained in the same formation; they are groups, subdivisions, or, as it is said by the geognosts of the school of Freiberg, members more or less developed of the same formation. (Freiesleben, *Kapf.*, t. i. p. 17., t. iii. p, 1.)

Notwithstanding the mixture of pelagic and fluviatile shells, which is sometimes observed at the contact of two formations of different origin, the name of *marine limestone*, or of *marine sandstone*, may be given to one of those formations, when we wish to derive the denomination of the rocks only from the species which constitute the greatest mass, and the middle of the beds. This terminology refers to a fact that belongs to the science of geogony, and to the ancient history of our planet: it decides (perhaps somewhat too much) on the alternation of fresh and salt-water. I do not deny the utility of the denominations *marine sandstone* or *marine limestone* for local descriptions; but, according to the principles which I propose following in this essay on formations, characterised according to the place which they occupy as the terms of a series, it appeared to me that I ought here carefully to avoid them. Are all the formations below the chalk, or even below the limestone with cerithia (calcaire grossier of the basin of Paris), without exception, marine limestone and marine sandstone? Or do the monitors and fish of the copper slate in the alpine limestone of Thuringia; the ichthyosauri of Mr. Home, placed below the oolites of Oxford and Bath in the lias of England, (which is represented on the Continent by a part of the Jura limestone); the crocodiles of Harfleur, buried in the clay with limestone beds above the oolites of Dive, and the limestone of Isigny, (consequently superior to the limestone of the Jura,) prove that below the chalk, between this formation and the red sandstone, there are small formations of fresh-water interposed in the great *marine formations?* Do not beds of coal with impressions of fern, beneath the red sandstone and the secondary porphyry, furnish an evident example of a very ancient formation not marine? These circumstances point out, in the present state of the science, how much caution should be used when we attempt, from characters that are merely zoological, to divide formations, the unity of which seemed certain from the alternation of the same beds, and from other phenomena of position. (Engelhard and Raumer, *Geogn. Vers.*, p. 125. 153.) This caution is so much the more necessary, since, according to the testimony of a mineralogist who has long deeply investigated this matter, M. Brongniart, there exists a sort of transition between the formation of marine limestone and that of the fresh-water gypsum which follows this limestone; and those two formations do not exhibit that abrupt separation which occurs in the same places between the chalk and the calcaire grossier, that is, between two marine formations. The same observer adds, "it cannot be doubted that the first beds of gypsum were deposited in a liquid analogous to the sea, while those that succeeded were deposited in a liquid analogous to fresh water." (*Geogn. Min.*, p. 168. 193.)

In stating the reasons which prevent me from generalising a terminology founded on the contrast between fresh-water and marine productions, I am far from disputing the existence of a fresh-water formation, superior to every other tertiary formation, and which contains only bulimi, limneæ, cyclostomæ, and potamides; recent observations have demonstrated that this formation is more generally spread, than was at first supposed; it is a new and last term to be added to the geognostic series. We owe the accurate knowledge of this fresh-water limestone to the useful labours of M. Brongniart. The phenomena presented by the fresh-water formations, the existence of which was formerly known only by the tufas of Thuringia, and by the ever-renowned *travertino* of the plains of Rome, (Reuss, *Geogn.*, t. ii. p. 642.; Buch, *Geogn. Beob.*, t. ii. p. 21—30.,) are connected in the most satisfactory manner with the admirable laws which M. Cuvier has noticed in the position of the bones of viviparous quadrupeds. (Brongniart, *Annales du Museum*, t. xv. p. 357. 581.; Cuvier, *Rech. sur les ossem. fossiles*, t. i. p. 54.)

The distinction between fluviatile and marine fossil shells is an object of the most delicate research; for it may happen, since the remains of organised bodies are detached with difficulty from the mass of siliceous limestone in which they are contained, that ampullaria may be mistaken for naticæ, and potamides for cerithia. In the concha family, we cannot with certainty separate cyclades and cyrenæ from venuses and lucinæ, but by an examination of the teeth of the hinge. The work undertaken by M. Ferussac, on land and fluviatile shells, will throw great light on this important object. Also when we think we perceive a genus of pelagic shells in the midst of a genus of fresh-water shells, we may enquire, whether in fact the same general types may not be found in lakes and seas. An example is already known of a true fluviatile mytilus; perhaps the ampullaria and corbulæ afford analogous compounds of marine and fresh-water forms. (See a Memoir of M. Valenciennes, inserted in my *Receuil d'Obs. de Zoologie et d'Anatomie Comparée*, t. ii. p. 218.)

From these general considerations on zoologic characters, and the study of fossil bodies, it follows, that notwithstanding the admirable labours of Camper, Blumenbach, and Sommering, the exact determination of species, and the examination of their relations with beds very recent and close to the chalk, does not date farther back than twenty-five years. I believe, that the study of fossil organic bodies, applied to all the other secondary and intermediate beds by geognosts, who consult at the same time the position and mineral composition of rocks, far from overthrowing the whole system of formations already established, will rather serve to support, improve, and complete its vast series. The geognostic science of formations may, no doubt, be investigated under very different points of view, according as we give a preference to the superposition of mineral masses, to their composition (that is, their chemical and mechanical analysis), or to the fossils which are contained in many of those masses; the whole of these are included in the science. The denominations, *geognosy of position*, or of *superposition*, *oryctognostic geognosy* (which considers the texture of the masses), and *geognosy of fossils*, designate, not branches of the same science, but various classes of relations which it is necessary to insulate, in order to study them more particularly. This unity of the science, and the vast field it comprehends, were well recognised by Werner, the founder of positive geognosy. Although he did not possess the necessary means for attaining a vigorous determination of fossil species, he never failed, in his course of lectures, to fix the attention of his pupils on the relations that exist between certain fossils and formations of different ages. I witnessed the high satisfaction which he felt, when M. de Schlottheim, one of the most distinguished geognosts of the school of Freiberg, began in 1792 to make those relations the principal object of his studies.

Positive geognosy has been enriched by all the discoveries that have been made on the mineral constitution of the globe, and furnishes valuable materials to another science, improperly called the *theory of the earth*, which comprehends the first history of the catastrophes of our planet. It reflects more light on that science than it receives in its turn; and without contesting the ancient fluidity or the softness of the stony beds, (a phenomenon proved by the fossil bodies, by the crystalline aspect of the masses, by the rolled pebbles, or the fragments imbedded in the transition and secondary rocks,) positive geognosy does not pronounce on the nature of the liquids in which it is said that the deposits were formed, those *waters of granite, porphyry,* and *gypsum,* which in hypothetic geology, are made to arrive tide by tide on the same point of the globe.

In this essay on formations, I have not indicated

the dip of the strata as a geognostic character No doubt, the discordance of two rocks, (Un gleichformigkeit der Leigerung,) that is, the want of parallelism in their direction and dip, is most frequently an evident proof of the independence of the formations; no doubt, the great inclination of the coal formation (coal measures), the red sandstone, and transition rocks, so justly contrasted in England by Mr. Buckland, with the horizontality of the magnesian limestone, red marl, lias, and all the more modern beds, is a phenomenon well worthy of attention; but in other regions of the earth, on the continent of Europe, and in equinoxial America, the alpine and Jura limestone are also highly inclined. In comprehending under the same point of view a vast extent of the globe, the Alps, the metalliferous mountains of Saxony, the Apennines, the Andes of New Grenada, and the Cordilleras of Mexico, we observe, that the inclination of the strata does not at all augment (as has been stated in some highly esteemed works,) according to the age of the formations. Sometimes, and in formations of considerable extent, there are beds almost horizontal among very ancient rocks: and farther, these phenomena have been observed rather among the primitive than the transition rocks; and in the former, rather among gneiss and stratified granites, than among clay-slates and mica-slates. It appears to me, that in general, those rocks that are most inclined are found (if we omit the beds very near to high chains of mountains,) between primitive mica-slate and the red sandstone. The horizontality of strata is very general, and strongly marked only above the chalk in tertiary formations; consequently, in masses of comparatively inconsiderable thickness.

This is not the place to examine the question, whether all inclined beds have been elevated or heaved up, as Stenon pretended in the year 1667, and which seems to be proved by the local phenomena of pebbles, or flattened fragments, placed parallel to the surfaces of inclined beds in the conglomerates of transition (grauwacke), and in the *nagelfluhe*; or, if it be possible that attractions which have acted at the same time on a great part of the surface of the globe have produced in our plains strata originally inclined, similar to the superposed and originally inclined laminæ which form the cleavage of a crystal. Certain sandstones (Nebra) exhibit a very regular parallelism in their thinnest layers, cutting at an angle of 20° to 35°, the fissures of horizontal or inclined stratification. Without attempting to solve these problems, I may be permitted to bring together some facts at the end of this introduction, that are connected with the study of positions. When, in the midst of a country not mountainous, or on table-lands uninterrupted by valleys, where the rock remains always visible, we travel during eight or ten leagues in a direction which cuts that of the beds at right angles, and find those beds (of transition clay-slate) parallel to each other, and almost equally inclined, from 50° to 60° to the north-west for instance, it is difficult to form an idea of a raising or lowering so uniform, and of the dimensions of the mountain or the hollow, which must be admitted to explain this inclination of the strata by a violent and simultaneous impulsion. In reasoning on the origin of inclined beds, we must distinguish two very different circumstances; their position in the proximity of a high chain of mountains, which is crossed by longitudinal or cross valleys, and their position at a distance from any chain of mountains, amidst plains or table-lands little elevated. In the first case, the effects of heaving up appear often incontestable, and the beds dip generally towards the chain; that is, on the northern declivity of the Alps towards the south, and on the southern declivity, but much less regularly, towards the north; (Buch, *in Schr. Nat. Focunde.*, 1809, p. 105. 109. 179. 181.; Bernouilli, *Schweiz. Miner.*, p. 25.) but at great distances from the chain, it appears to influence only the direction, but not the dip of the beds.

Since the year 1792 I have been attentive to this parallelism, or rather to this *loxodromism* of beds. Residing on mountains of stratified rocks, where this phenomenon is constant, examining the direction and dip of primitive and transition beds, from the coast of Genoa across the chain of the Bochetta, the plains of Lombardy, the Alps of St. Gothard, the table-land of Swabia, the mountains of Baireuth, and the plains of Northern Germany,

had the advantage of making my first mineralogical studies, under a great master, and who on a vast extent of territory (between 28° and 71° of latitude) has collected valuable materials for geognosy, the history of the atmosphere, and the geography of plants. In the course of my labours I have made use of many unpublished notes which this naturalist had the kindness to give me on the crystalline structure of the trachytes which I brought from the Cordilleras, and on the order of formations in Switzerland, England, Scotland, Tuscany, and the vicinity of Rome. I have also had the advantage of consulting him during his visits to Paris, on what appeared to me to be doubtful in the position of formations. All the observations relative to Hungary are taken from the *Voyage Minéralogique* of M. Beudant, which will soon be published; and in which the subject of position is, for the most part, treated in a superior manner. My countryman, M. Charpentier, director of the salt-works in Switzerland, kindly communicated to me his excellent description of the Pyrenees, the most complete work which we possess on a great chain of mountains. Much of the information on the porphyries of Europe is taken from a sketch that I wrote in some degree under the inspection of M. Werner, when that celebrated man came, for several days, from Carlsbad to Vienna, (in 1811,) in order to converse with me on the geognostic constitution of the Cordilleras of the Andes, and of Mexico. To give a public testimony of gratitude to those whose memory is dear to us is a duty which it is grateful to fulfil. I have not reaped all the advantage which I wished from the important labours of MM. Maculloch, Jameson, Weaver, Berger, and other members of the *geological* and *Wernerian* societies in England, because I feared to pronounce on the identity of the formations of a country with which I was unacquainted, north of the mountains of Derbyshire, and which is at present explored with so much zeal and success.

By indicating for each formation the names of some of the places where they are found, (what the botanists term *the habitats*,) I have in no wise any pretension to extending the domain of mineralogical geography; I only endeavoured to give examples of position already well observed. Those examples are not always chosen amidst countries which, from the descriptions of celebrated geognosts, are become in some sort *classical*. It was sometimes proper to name places in the other hemisphere, that are not found on any of our maps. Allemont, Dudley, Cap de Gates, Mansfield, and Œningue, are better known to mineralogists than the great metalliferous provinces of Antioquia, Guamalies, and Zacatecas. To facilitate this kind of research, I have often added between two parentheses, some geographical notices; for instance, Quindiu (New Grenada), Tiscan (Andes of Quito), Tomependa (plains of the Amazon). With the indication of the places where certain formations predominate, I have endeavoured to state the whole order of superposition which has been observed with some degree of certainty on very distant points; for instance, in the Cordilleras of the Andes, in Norway, Germany, England, Holland, and at the Caucasus. Those descriptions of profiles or vertical *sections* which furnish the materials so long desired for the construction of a *geognostic atlas*, constitute, it may be said, the evidence that establishes a general tabular arrangement of rocks; for geognosy, when occupied with the series of formations, is to mineralogical geography, what *comparative hydrography* is to the topography of great rivers separately traced. It is from the intimate knowledge of the influence exerted by inequalities of the surface, the melting of snow, periodical rains and tides, on the swiftness, the sinuosities, the contractions, the bifurcations and the form of the mouths of the Danube, of the Nile, of the Ganges, and of the Amazon, that we form a general theory of rivers, or rather a *system of empirical laws*, that includes all that is common and analogous, in local and partial phenomena. (See some elements of this comparative hydrography, in my *Relat. Histor.* tom. ii. p. 517—526., and 657—664.) The *geognosy of formations* also offers *empirical* laws, which have been abstracted from a great number of particular cases: founded on mineralogic geography, it differs from it essentially, and this difference between abstraction and individual observation may, among geognosts who have only seen one country, become the cause of an erroneous judgment on the accuracy of a general table of formations.

The physical sciences depend in a great measure on inductions, and the more complete those inductions become, the more it is necessary to exclude the local circumstances that accompany every phenomenon in announcing general laws. The history of geognosy justifies this assertion. Werner, in creating geognostic science, has perceived with an admirable sagacity all the relations under which we should view the independence of the primitive, transition, and secondary formations. He has shown what we ought to observe, — what it is important to know; he has prepared, and foreseen in some degree, a part of the discoveries with which, through him, geognosy has been enriched in countries which he could not visit. As formations do not follow the variations of latitude and climate, and phenomena, observed perhaps for the first time in the Himalaya, or the Andes, are found again, and often with an association of circumstances that seem to be entirely accidental, in Germany, Scotland, or the Pyrenees, a very small portion of the globe, a territory of some square leagues in which nature has assembled many formations, may, (like a true *microcosm* of the ancient philosophers), give rise, in the mind of an excellent observer, to very accurate ideas on the fundamental truths of geognosy. In fact, the first views of Werner, even those which that illustrious man had formed before the year 1790, possessed a justness that is still remarkable. The learned of every country, even those who show no predilection for the school of Freiberg, have preserved them as the basis of geognostic classifications; and yet what was known, however, in 1790, of primitive, transition, and secondary formations, was founded almost entirely on Thuringia, on the metalliferous mountains of Saxony, and those of the Harz, on an extent of country not 75 leagues in length. The memorable labours of Dolomieu, and Saussure's descriptions of the Alps, were consulted, but these could not have influenced in a great degree the labours of Werner. Saussure has, no doubt, given inimitable models of accuracy in the topography of every summit, and every valley; but that intrepid traveller, struck with the complicated nature of the phenomena of superposition, and the apparent disorder which prevails every where in the interior of the high alpine chains, was not much induced to occupy himself with general ideas on the geognostic structure of an entire region. In the first period of the science, the *type of formations* was founded on a small number of observations, and resembled too much the description of the places where it originated. The mineral masses which in other countries are but subordinate or accidental beds, were mistaken for independent formations; the existence of those formations that are important in equatorial America, and in the north and west of Europe, was unknown; the relative antiquity of porphyries, syenites, and euphotides, was misunderstood; the history of the more recent beds was not completed by a rigorous determination of the organised fossil bodies which they contain; the position of basalts, phonolites (porphyr-schiefer), and dolerites, which had been long confounded with trap-greenstone, was studied with great precision, but even the possibility of their igneous origin was denied; because, in the country where modern geognosy had its birth, the observer was surrounded only by some remains of volcanic formations, and could not examine the relations that exist between the trachytes (trap-porphyr), the basalts, the more modern lavas, scoriæ, and pumice. If Werner's list of formations, notwithstanding the books which he consulted, and the surprising sagacity with which he discerned the truth in the frequently confused narratives of travellers, still remained incomplete, he felt no regret in seeing his labours improved by other hands. He was the first who taught the art of observing and distinguishing the formations; and it is by the further application of that art that geognosy has become a positive science. Conscious that his real glory was rather founded on the discovery of the principles of the science, and on the means of research, rather than on the results obtained at a particular epoch. Werner showed no less regard for such of his pupils as differed from him on the subject of the relative age, and the origin of some of the formations. It is only in extending our observation to a greater part of the globe, that the type of formations can be enlarged and simplified. It has thus been made to correspond better with the geognostic constitution of continents considered in a general point of view.

We now know, with considerable accuracy, the relative position of many formations. 1°. In the *ancient continent:* in Great Britain, in the north of France, Belgium, Norway, Sweden, and Finland; in Germany, Hungary, Switzerland; in the Pyrenees, Lombardy, Tuscany, and the vicinity of Rome; in the Crimea and Caucasus; (lat. 42° —71° nor.; long. 40° or.—12° oc.). 2°. In the *new continent:* in the United States of America, between Virginia and the Lake Ontario, (lat. 36° —43° nor.; long. oc. 78°—86°); at Mexico, between Veracruz, Acapulco, and Guanaxuato, (lat. 16° 50′—21° 1′ nor.; long. oc. 98° 29′—103° 22′); in the island of Cuba (lat. 23° 9′ nor.); in the United Provinces of Venezuela, between the coast of Paria, Portocabello, the Upper Oroonoko, and San Carlos del Rio Negro; in the Andes of New Grenada, Popayan, Pasto, Quito, and Peru; in the valley of the river of the Amazons, and on the coast of the South Sea, (lat. 10° 27′ nor.—12° 2′ austr.; long. oc. 66° 15′—82° 16′); at Brazil, between Rio Janeiro and the western limit of the province Minas Geraes (lat. 18° 23′ austr.; long. oc. 45°—49°). In proportion as we ascend to more general ideas, the table of formations, in becoming more extensive, and (we venture to believe) more exact, is less satisfactory to those who expect to find in it a stronger expression of the individual features, and the local physiognomy of their canton. But those individual features, that local physiognomy, cannot be preserved there but as simple variations of a general type; as particular modifications of the great laws of geognostic position. However incomplete may yet be the knowledge of those laws, a great step is already made in these researches by having acquired, from the united labours of our contemporaries, the certainty that constant and immutable laws do exist amidst the conflict of local perturbations.

PRIMITIVE FORMATIONS.

The most ancient formations of primitive rocks that have come under our observation, are, in some regions of the globe, *granite* (a formation in which granite alternates with no other rock); and in other regions, *gneiss-granite* (a granitic formation, in which beds of granite alternate with beds of gneiss). It would be difficult to mention a granite that geognosts unanimously consider as anterior to every other rock; but this uncertainty belongs to the nature of the subject, and to the idea that we form of the relative age and the superposition of rocks. We may assure ourselves by observation, that the granite of St. Gothard reposes on mica-slate, and that that of Kielwig in Norway rests on clay-slate. But how can we demonstrate a negative fact? How can we prove that beneath a granite, which is said to be of the first formation, gneiss, or some other primitive rock, may not be again found? In tracing a sketch of the knowledge we have acquired of the superposition of rocks, we ought to abstain from pronouncing with confidence on the first stage of the geognostic edifice. Thus (for it is with time as with space), through the long continued migrations of tribes, the historian is unable to ascertain with certainty, who were the first inhabitants of a country.

I. PRIMITIVE GRANITE.

§ 1. Granite that does not alternate with gneiss. Many doubts not ill founded having lately been raised with respect to the antiquity of many formations of granite, this first of primitive rocks can only be designated by negative characters. It appears to me, that in both hemispheres, particularly in the new world, granite is most ancient, when it is not stratified, is richer in quartz, and less abundant in mica. In lofty chains of mountains (in the Alps of Switzerland and the Cordilleras of the Andes, between Loxa and Zaulaca,) granite, by the abundance and uniform direction of the scales of mica, has a tendency to become lamellar; while the granites that pierce through the vegetable soil in the plains generally exhibit, by their more uniformly granular texture, a more striking contrast with gneiss. The coarseness of the grain, the regularity in the crystallisation of the constituent parts, and the red or white colour of the feldspar, are phenomena well worthy of attention, if we consider the great masses of a rock, and omit the subordinate beds of small-grained granite that are found amidst coarse-grained granite, and *vice versâ*. These phenomena may mark the relative age of a formation in a circumscribed extent of country; but general characters applicable to an entire continent cannot be deduced in this manner. In the Cordilleras, the small-grained granite, with white and yellowish-white feldspar, appeared to me the most ancient. The absence, I do not say of tourmaline and of rutile, but of disseminated hornblende, steatite, garnets, epidote, actinolite, tin, and specular iron, replacing mica (Gottesgabe in the Upper Palatinate); the want of subordinate heterogeneous beds (greenstone, granular limestone) and nodules with very small grains and highly micaceous, which are of contemporaneous formation, and seem imbedded in the principal mass; finally, the want of stratification in the lower beds, and a more porphyroid structure, appear to characterise the granites of the first formation, (western coast of equinoxial America, Cascas, Santa, and Guarmay in Lower Peru; the banks of Cumbeima, Ibagné, Quilichao, and Caloto in the Andes of New Grenada). The granites of the cataracts of the Oroonoko, and the mountains of Parima, contain, like those of the Pyrenees and Upper Egypt, some beds in which insulated crystals of hornblende are found; these rocks probably belong to a more recent epoch than the granite of Lower Peru. Although the most ancient granites, in general, have no subordinate beds of primitive limestone, calcareous matter begins to appear in the primitive mountains (I dare not say at the first period of the world) in feldspar, and perhaps, in the tourmalines. This quantity of lime afterwards augments by the addition of hornblende in the syenitic beds which characterise the most modern granites.

PRIMITIVE GRANITE AND GNEISS.

§ 2. In this formation, so well characterised by M. Raumer, very distinct beds of granite and gneiss occur nearly contemporary, and alternating with one another. It sometimes reposes (Reisengebirge) immediately on the preceding formation, at other times (south-east of Riobamba, in the kingdom of Quito,) it is the most ancient of the visible rocks. This periodical return of heterogeneous beds, is found particularly in transition formations; for instance, in those of porphyry and syenite, and syenite and greenstone. We should distinguish from the formation of granite and gneiss, the granites of the shore of Venezuela, the beds of which pass often and insensibly to gneiss, and gneiss which passes to granite (southern declivity of the Jungfrau and the Titlis). The beds subordinate to granite and gneiss are mica-slate containing granular limestone, hornblende, chlorite-slate, and whitestone (weistein).

STANNIFEROUS GRANITE.

§ 3. In this rock the constituent parts are generally very much disintegrated, the feldspar passing to kaolin (Carlsbad, on the road from Eibenstock to Johann-Georgenstadt; and probably also, according to M. Bonnard, the granites of the department of Haute-Vienne). It will, perhaps, be hereafter discovered that several of those stanniferous rocks are of a date still more recent, and that they must be placed among granites posterior to gneiss, and anterior to mica-slate. It would appear that new characters have been found in the granites of the Fichtelgebirge, in Franconia, which are not only very regularly stratified, but also contain beds of *greenstone* (primitive diabase paterlestein). I am not acquainted with the alpine formation of stanniferous granite in the Andes; the granite that constitutes the summits of the Cordilleras is almost always covered by transition porphyry and trachyte.

WHITESTONE (WEISTEIN, EURITE) WITH SERPENTINE.

§ 4. Whitestone (eurite), in which compact feldspar predominates, (north-east part of Erzgebirge,) rests on ancient granite. It is covered by gneiss, and sometimes by mica-slate (Hartha), or by a primitive slate into which (Hermsdorf, Döbeln) whitestone appears to pass insensibly. *Subordinate beds:* granite, sometimes with very large grains (Penig); sometimes fine-grained passing to whitestone, and containing lepidolite and lamellar parenthine; serpentine (Waldheim). Whitestone, in which garnets and syenite are sometimes imbedded, is, in Saxony, according to the observations of MM. Pusch, Raumer, and Mohs, an independent formation anterior to gneiss, and not a subordinate bed; in Silesia, (Engelsberg, near Zobten Weiseritz, near Schweidnitz), it only forms beds in primitive granite and gneiss. There is nothing in this phenomenon that should astonish the geognost. Mica-slate, gneiss, and porphyry, are found both as independent rocks and as subordinate beds. The serpentine of Buenavista, in the mountains of Higuerote, west of the Caracas, belongs properly to talcose gneiss; but it appears, that in the same group of mountains there is also serpentine joined to a whitestone, which is superposed on the formation of granite and gneiss. The serpentine of whitestone is the most ancient of the very small-grained euphotide rocks, which, it may be said, pass through all the following formations, as far as the upper limit of the transition series.

II. PRIMITIVE GNEISS.

§ 5. We distinguish this formation of gneiss (Freiberg; Lyons, table-land between Autun and the mountain of Aussi; Arnsberg, in the Riesengebirge; Lödingen, in Norway; the Grampians, in Scotland;) which contains subordinate beds of mica-slate, from a no less important formation of gneiss and mica-slate, in which the beds of gneiss alternate with beds of mica-slate. Gneiss, according to MM. de Buch and Haussmann, is the prevailing rock in Scandinavia, where the ancient granite (anterior to gneiss) is scarcely any where visible. The subordinate beds of gneiss are much varied and numerous; they are, however, much less so when the gneiss does not pass to mica-slate. We shall here enumerate only the most remarkable beds: quartz, often containing garnets; feldspar more or less decomposed, and without potash; porphyry, generally reddish, with a petro-siliceous base, containing feldspar, quartz, and mica, (lager-porphyr of Halsbrücke, Ober-Frauendorf, and Liebstadt); granular limestone but rarely, (road of the Simplon, mine of Kurprinz, near Freiberg); common garnet, mixed with granular limestone, blende, and oxidulated iron, (Schwarzenberg); mica-slate, (Bergen, in Norway); syenite, (Burkersdorf, in Silesia); granite, with decomposed feldspar, but not stanniferous; serpentine (ophiolite), forming, according to M. Cordier, a bed of immense extent, in the departments of Haute-Vienne, Lot, and Aveyron; slaty amphibolite or hornblende-slate; greenstone mixed with magnetic iron, (Taberg, near Jonköping); zircon, zoïsite, and menachanite, (Priockterhalt, in Carinthia); magnetic iron, in beds from twenty to thirty toises thick, often mixed with granular limestone, ichthyophthalmite, spodumene, tremolite, amianthus, actinolite, and bitumen, (Danemora, Gellivara, and Kinsivara, Sweden and Lapland); pegmatite, (Loch-Laggan, in Scotland); gneiss, containing angular masses of gneiss of a different texture from that of the principal rock, (Rostenberg, in Norway). This latter phenomenon (the effect of a contemporary crystallisation ?) is much more analogous to the granite of Greiffenstein, in Saxony, and of the Pic Quairat, in the Pyrenees, than to the transition gneiss, containing the pudding-stones of Valorsine. The great formation of primitive gneiss, rich in the ores of silver and gold, in Germany, some parts of France, Greece, and Asia Minor, has been long considered as the most argentiferous rock of the globe. We now know, from researches made in both Americas, and in Hungary, that the great mass of precious metals which circulate in the two continents has been procured from formations much later than gneiss, and every other primitive formation, that they come from transition rocks, syenitic porphyries, and even trachytes. Gneiss, little metalliferous in the equinoxial part of the new world, occurs on a greater extent of country in the mountains that run from east to west, (chain of the shore of Caracas, cape Codera, and the isles of the lake of Tacarigua; Oroonoko, Sierra of Parime;) and in the low regions distant from the chain of the Andes (east of the mountains of Brazil), than in the elevated crest of the chain itself. I saw no gneiss (at the Silla of Caracas, and at the passage of the Andes of Quindiu) higher than from 1300 to 1400 toises above the level of the ocean. On the crest of the Cordilleras, between Ibague and Carthago (New Grenada or Cundinamanca) as well as at the Paramo of Chulucanas, in descending towards the Amazon, a granite of new formation covers the gneiss at 1800 toises high. If gneiss, mica-slate, and a granite of second formation, constitute the most lofty summits in the mountains of Europe; the most elevated summits of the Andes, on the contrary, exhibit only immense accumulations of trachytic rocks. In following the same chain, the same line of mountains, we find the low regions of granite-gneiss, and gneiss-mica-slate (province of Oaxaca in New Spain, where gneiss is auriferous; primitive groups of Quindiu, Almaguer, Guamote, at the south of the Chimborazo; Sargura and Loxa, in the Andes of Peru;) alternating with the elevated regions of trachytes (1000 to 3300 toises). These latter formations, produced or modified by fire, undoubtedly cover granite and gneiss, and that sometimes immediately without the porphyritic formations of transition being interposed. However, where I could see the trachytes of the kingdom of Quito (volcano of Tunguragua, ravine of Rio Puela, near Penipe,) reposing on a greenish micaceous slate filled with garnets, and covering in their turn a granite, rather syenitic, with quartz and mica (black ?), this superposition takes place only at the inconsiderable height of 1240 toises. It results in general from my barometric levelling of the Cordilleras, that in all that region of the tropics, ancient granite and gneiss (which must not be confounded with syenitic and granitic transition rocks) scarcely rise above the height of the summit of the Pyrenees. All the superposed masses of primitive rocks, which pass the limit of perpetual snow (2300—2460 toises), and which give the Cordilleras their character of grandeur and majesty, are not owing in general, either to primitive formations, or calcareous rocks, (it is only the alpine limestone of the table-lands of Gualgayoc and Guancavelica, which occurs at 2100 and 2500 toises,) but to trachyte-porphyries, dolerites, and phonolites. (We do not yet know of what rocks the summits of the Himalaya, those peaks which have been recently measured by Mr. Webb, are composed.) The gneiss of the Cordilleras abounds much more than the mica-slate in subordinate beds of granular limestone (micaceous and filled with pyrites). In equinoxial America, therefore, as at the most northern extremity of Europe, and in the Pyrenees, garnet is most frequent in gneiss, and this latter rock only ceases in general to contain garnets where it passes to mica-slate (mountain of Avila, near Caracas). A true gneiss, without garnets, occurs, however, at the west of Mariquita, between Rio Quamo and the mines of Sana (New Grenada). In Brazil, according to the observations of M. d'Eschwege, tin (zinnstein) is disseminated, not in granite, but in gneiss (banks of Rio Paraopeba, near the Villa-Ricca).

Between the two great formations of primitive gneis and mica-slate, we shall place several parallel formations:—

GNEISS AND MICA-SLATE;	PRIMITIVE SYENITE?
GRANITE POSTERIOR TO GNEISS, AND ANTERIOR TO MICA-SLATE.	PRIMITIVE SERPENTINE?
	GRANULAR LIMESTONE.

Two of those formations are perhaps no less doubtful than primitive porphyry, considered as an independent formation.

GNEISS AND MICA-SLATE.

§ 6. Beds of gneiss alternate with beds of mica-slate, in the same manner as gneiss, in the formation (§ 2.), alternates with granite. They are not rocks that pass into each other, but alternating beds very distinct, (Neisbach and Janersberg in Silesia; Weltersdorf near Schisberberg, in Saxony). In the Cordilleras of America, and perhaps for the most part in the great chains of mountains of the ancient continent, (as the illustrious Dolomieu made me observe in Switzerland, in the year 1795,) the *mixed* formations and *periodical alternations* of gneiss and granite, and of gneiss and mica-slate, are much more frequent than the *simple* formations of granite, gneiss, and mica-slate. The independent formation of gneiss-mica-slate reposes, sometimes, on the formation of gneiss (§ 5.), sometimes immediately on the most ancient granite (§ 1.). In the latter case, it should be considered as a parallel formation to gneiss.

Subordinate beds: granular limestone, hornblende-slate, greenstone, serpentine, and clay-slate with actinolite. These subordinate beds are several times repeated; for in all the formations of *periodical alternation*, either primitive or transition, (granite and gneiss, gneiss and mica-slate, syenite and greenstone, porphyry and syenite, porphyry and grauwacke, black limestone and transition schist,) the periodical return of masses, extends even to the subordinate beds. This great geological law manifests itself in the whole Cordilleras of the Andes, particularly, in the mountains situated at the south and south-east of the volcano of Tunguragua, at Condorasto, and at Paramo del Hatilla, where (what is very rare in that region) gneiss-mica-slate rises to more than 2000 toises, and contains veins of silver formerly celebrated (weisgültigerz, and sprödgluserz, white silver, and brittle silver ore). These metalliferous gneiss-mica-slates of Condorasto and Pomallacta, are hid towards the south beneath the formations of a trachytic porphyry of the Andes of Assuay; they re-appear (at 1700 toises high) between the ruins of the palace of the Inca (Ingapilca) and the farm of Turche, and are again concealed beneath the sandstone of Cuenca. The forests of Quinquina, west of Loxa, cover also those mountains of gneiss, alternating with mica-slate. In the passage of the Andes of Quindiu, between the basins of the Rio Cauca and Rio Magdalena, the formation of gneiss-mica-slate reposes (above the station of the Palmilla,) immediately on ancient granite. It attains an enormous thickness, in rising towards the Paramo of San-Juan. The beds of mica-slate alternating with gneiss are here always destitute of garnets; they are traversed at the Valle del Moral (at 1065 toises high) by veins filled with sulphur, exhaling sulphureous vapours, the temperature of which rises to 48° cent., the atmospheric air being at 20°. This phenomenon is the more remarkable as at the south of the equator, in the celebrated *moun-*

tain of sulphur of Tiscan, I found sulphur in quartz, as a subordinate bed in primitive mica-slate. The gneiss of Quindiu contains disseminated garnets and beds of decomposed kaolin. In the littoral chain of the Caracas, between Turiamo and Villa de Cura, the granite-gneiss and gneiss-mica-slate formations occupy, in a direction perpendicular to the axis of the chain, a space of ten leagues broad; the gneiss-mica-slate is concealed, towards the Llanos of Venezuela, beneath green schists of transition. Near Guayra, at Cape Blanc, this formation contains subordinate beds of chlorite-slate (with garnets and magnetic sand), and hornblende-slate and greenstone, mixed with quartz and pyrites. On the coast of Brazil, where several primitive chains run parallel with the Andes of Peru and Chili, in the direction of a meridian, beds of granite, gneiss, and mica-slate, constitute a single formation, and alternate in a periodical series, (Ilha Grande at the south of Rio-Janeiro, near Villa d'Angro dos Reis, according to M. Eschwege). The three rocks are there contemporary, like the syenites, which alternate periodically, either with clay-slate, or transition greenstone.

GRANITE POSTERIOR TO GNEISS, AND ANTERIOR TO MICA-SLATE.

I here unite several formations of granite nearly parallel, placed between gneiss and mica-slate, such as stanniferous granite (hyalomicte graisen) of

Zinnwald and Altenberg in Saxony, which appears to repose on gneiss, and abounds in black tourmalines; the greater part of the pegmatites or graphic-granites (schriftgranite), which contains lepidolite, (Rosena, in Moravia); granites with epidote; granites with subordinate beds of white-stone, or eurite, (Reichenstein, in Silesia); granites with steatite and chlorite, often containing disseminated hornblende, and assuming the aspect of a syenite, or a chlorite-slate (protogine of Mont Blanc, and of almost the whole chain of the Alps, between Mont Cenis and St. Gothard; probably also the rock of Rehberg in the Harz); the granites of the Pyrenees, so well studied by M. Charpentier, and containing numerous beds of gneiss, mica-slate, and granular limestone. The granites of Altenberg, perhaps, belong (such is the opinion of M. Beudant) to the lower beds of transition porphyry; perhaps the granites of the Pyrenees, which enclose masses of greenstone (primitive diabase), are even posterior to the great formation of mica-slate (§ 11.), as also the stanniferous granites of the Fichtelberg, which contain greenstone, (Ochenkopf, Schneeberg, in Franconia), and which we have indicated provisionally at § 3. I have the same doubt respecting many granites that abound in argentiferous veins; particularly the granites with garnets, and the porphyroid granites (with very large crystals of red and white feldspar), which are often stratified with as much regularity as secondary limestone. I would not have noticed the masses of tin of Geyer and Schlackenwald, because the granites in which they are contained are only beds in gneiss and mica-slate; they are not true rocks, and independent formations, like the granites of Carlsbad of the Fichtelgebirge. In equinoxial America we may, with some probability, connect with the formation of granite posterior to gneiss and anterior to mica-slate, the granites of the western declivity of the Cordilleras of Mexico, (table land of Papagallo and Moxonera,) which are of a porphyroid structure, or divided into balls, with concentric layers. They enclose syenitic beds, connected with veins of bassanite (compact urgrünstein). I have seen them regularly stratified in beds, seven or eight inches thick, and having, not the same inclination, but the same direction with the strata of transition porphyry and superposed alpine limestone. We do not know, indeed, the rocks that cover this Mexican formation of granite, on which, however, all the other rocks of Mexico are placed; but the great characters of composition and structure which it exhibits, and its analogy with other stratified granites of the high Andes of Peru, lead me to believe that it is of a more recent date than the formation § 1. To the granite anterior to mica-slate, but posterior to gneiss, belong more certainly that of the Garita del Paramo, at the foot of the extinguished volcano of Tolimo (Andes of Quindiu); that of Silla de Caracas; the granites, very regularly stratified (without passing to gneiss), of Los Trincheras, in the littoral chain of Venezuela; the granites of the extended group of the mountains of Parime, which are either regularly stratified, (strait of Baraguan, valley of the Lower Oroonoko), or pass to pegmatite, (Esmeralda and the confluence of the Uramu, Upper Oroonoko), or hornblende rocks (amphibolites), (cataracts of Atures). In this vast granitiferous group of Sierra Parime, which separates the basin of the Oroonoko from that of the Amazon, some phenomena are observed, which are also seen in Finland and Norway. No other mineral mass appears than the granite rock. In going round the Sierra Parime, at the north, the west, and the south, I observed, with the exception of some small masses of sandstone, a total absence of secondary formations even of rocks posterior to a granite of newer formation. This granite, and the gneiss that supports it, form (where little plains separate the mountains from each other, amidst forests of vigorous vegetation) beds of bare rock, destitute of soil, of more than 250,000 square toises in extent, and scarcely rising three or four inches above the surrounding soil. I notice, as granites of new formation, in the southern hemisphere, the rock of Pareton, (eastern declivity of the Andes of Peru, between Guamcabamba and the River of the Amazons), where the steatitic granite passes to protogine; the granite of Paramo, of Pata Grande, and Nunaguava, stratified and destitute of hornblende; the rock of Yanta, stratified like the granite of Ochsenkopf, in Franconia, concealed beneath the mica-slate of Gualtaquillo and Aipata, and containing disseminated crystals of hornblende, without passing to true syenite, (Cordilleras of Gueringa, at the west of Guamcabamba). We see by these examples, that in the Andes as well as in the Alps, particularly at a considerable height, a granite rock covers primitive gneiss. It may be asked, if the primitive greenstone, which forms beds in the formations §§ 3. 5, 6, 7. sometimes contains, as several geognosts assert, not only hornblende mixed with compact feldspar, but pyroxene also. M. de Charpentier saw this latter substance in large masses in the primitive limestone of the Pyrenees. There is also pyroxene-coccolite in the urgrünstein of the Lake Champlain; I saw true pyroxenes, identical with those of the trachytes and some transition porphyries of Quito, only in the transition greenstone and transition mandelstein of Parapara (mountains of Venezuela).

PRIMITIVE SYENITE?

§ 8. The greater part of the syenites of the ancient and the new continent, considered formerly as independent rocks of primitive formation, are either granites with hornblende; that is, beds subordinate to granite, §§ 7. and 11., (Syene, now Philæ, or even the first cataract of Upper Egypt, which are in gneiss; Atures, or the cataracts of Oronooko; valley of Macara and Gualtaquillo, and the eastern declivity of the Andes of Peru); or transition formations (Mount Sinaï, according to the interesting observations of M. Roziere; valley of Plauen, near Dresden; Guanaxuato, in Mexico,) intimately connected with porphyries, greenstones, and transition clay-slates. Some true syenites appear to me, however, to have no trace of this connection; they perhaps constitute independent primitive formations; such are the syenite (a great deal of reddish lamellar feldspar, little hornblende, scarcely any quartz, no mica, no titaneous iron) of Cerro Munchique (central Cordillera of the Andes of Popayan, at the east of the farm of Ciscabel, situated on gneiss, and partly (?) covered by primitive mica-slate); the syenite of Paramo de Yamoca (eastern declivity of the Andes of Peru, near the Indian villages of Colascy and Chontaly), placed on the granite of Zaulaca, and covered by the slate of the lake of Hacatacumba. As that slate supports, in its turn, a green transition porphyry, and that porphyry supports a blackish-grey shelly limestone, (San Felipe, province of Jaen de Bracamoros), it remains doubtful, whether the syenite of Yamoca and the slate of Hacatacumba be not also transition rocks; and, consequently, newer than the syenites of Cerro Munchique, in the Andes of Popayan. Are the syenites, composed of white feldspar and green hornblende, at the foot of Mont-Blanc, (Cormayeux,) and also the syenites of Biela connected with euphotides, to be considered as primitive formations?

PRIMITIVE SERPENTINE?

§ 9. The great formations of euphotide (gabbro, or serpentine rocks) are posterior to primitive clay-slate, and belong partly to transition rocks. The small formation which we here designate is analogous to that of Zœblitz, in Saxony; it reposes on gneiss, and is not covered by any other rock. The serpentine in South America, (without diallage metalloïde, but with garnets,) of the mountains of Higuerote (near San Pedro, between the town of Caracas and the valleys of Aragua), appears analogous to that of Saxony. It reposes on the talcose gneiss of Buenavista, that passes (which is rare in those countries) to a garnetiferous mica-slate. As we see, however, no rock superposed on those serpentines, their age remains a little doubtful. What appears to me to prove the antiquity of the serpentines of Higuerote, is, that before appearing as a particular and independent formation, they are seen as subordinate beds to mica-slate-gneiss, nearly in the same manner as the serpentines of the valley of Aoste.

PRIMITIVE LIMESTONE.

§ 10. Does there exist an independent formation of granular limestone among primitive rocks? Or, are all those granular limestones, which have hitherto been very generally admitted, only beds subordinate to gneiss, mica-slate, granites of new formation, and clay-slate. In the Pyrenees (Valley de Vicdessos,) M. Charpentier considers the granular limestone, sometimes blackish, mixed with graphite, and containing great masses of pyroxene (Iherzolite, augitfels,) and beds of greenstone, as an extended and independent formation. This authority has, no doubt, great weight. At the south of the equator, on the table-land of Quito (at Cebollar, and on the banks of the Rio Machangara, near Cuença; Portete, in the Llano of Tarqui,) we find, on mica-slate (of Guasunta and Cañar,) a coarse grained white limestone, resembling the finest Carara marble, and alternating with limestone beds that are almost compact, striped, and so translucid, that they are used in chapels and convents as window-glass. I long regarded this granular limestone of Cuença, void of petrifaction, as a primitive and independent formation; but it is only covered by the red sandstone of Nabou; and an analogous formation, (Tolonta near Chillo,) placed amidst transition trachytes and porphyries, renders the age of the formation of Cuença extremely doubtful. The beds of primitive limestone subordinate to rocks of granite-gneiss, are far more rare in equinoxial America than in the Pyrenees and the Alps. In examining with care the gneiss-granite of Parime, between the 2d and 8th degrees of north latitude, I did not find one of those beds.

III. PRIMITIVE MICA-SLATE.

§ 11. Mica-slate (schiste micacé, glimmer-schiefer) reposes most frequently on gneiss, at other times immediately on granite, § 1.), with which it first begins to alternate (Schneeberg in Saxony, Minas Geraes in Brazil) before it appears as an independent formation. It is distinguished from gneiss, when the two rocks are strongly characterised (which is far more rarely the case in the high chain of the Alps, and the Cordilleras of Peru, than in the plains) by the aggregation of the mica, which in mica-slate presents a continued surface. Of all the primitive formations this is the most developed in central Europe, and exhibits the greatest variety of subordinate beds. The beds become more heterogeneous in proportion as this rock becomes more distinct from granite. The mica-slates of the Pyrenees, which are considered as being decidedly primitive, often contain chiastolite; and that substance sometimes extends as far as the interposed beds of clay-slate and granular limestone. Beds subordinate to mica-slate: chlorite-slate (chlorit-schiefer with garnets); a mixture of mica-slate and granular limestone, (Splügen, between Glaris and Chiavenna; pic du midi of Tarbes, in the Pyrenees); clay-slate; granular limestone, and dolomite with tremolite (grammatite), epidote, talc, tourmaline, lepidolite, hornblende, corundum, and magnetic iron; granular limestone containing quartz, (Pyrenees);

dolomite mixed with primitive gypsum, (passage of Splügen in the Alps); slates and micaceous quartz (gestellstein); greenstone and greenstone-slate; granular and schistose diabase, (Montãna de Avila, Cabo Blanco near Caracas); blackish-green compact feldspar (dichter grünstein); pot-stone, (*topfstein*, Ursern); talc-slate (talk-schiefer) with garnets, syenite, tourmaline, and actinolite; pure serpentine, (Sillthal in the Tyrol); serpentine mixed wit hgranular limestone, *verde antico* (mountains of Caramania; Reichenstein, Rörsdorf, and Rothzeche in Silesia); hornblende-slate, (Saint Pierre, south of the Great St. Bernard); common hornblende in great masses, (Schönberg in the Tyrol); syenite, (Mittlewald in the Tyrol); beds of garnets with oxidulated iron, (Braunsberg near Freiberg, Frauenberg near Ehrenfriedrichsdorf, in Saxony); garnet with omphacite-pyroxene, and hornblende, (Gefrees and Schwarzenbach, country of Bareuth, Sanalpe in Carinthia); actinolite, garnets, and syenite; fluate of lime, (Messersdorf); beds of mica-slate containing masses of gneiss perhaps of a contemporary formation, (Tuffle in Norway); beds several feet in thickness, composed of an intimate mixture of compact feldspar, quartz, and mica, (Külstad near Drontheim, in Norway); mica-slate with black and carburetted mica, (Sneehättan in Norway; Huffiner in the Valais). I do not cite the gypsum of Val Canaria, near Airolo, which M. Freiesleben and myself, in 1795, believed to be a primitive formation subordinate to mica-slate; but which MM. Brochant and Beudant (who studied it separately with care) determined to be transition gypsum superposed on mica-slate. Mica-slate often contains hornblende disseminated through its mass, (Salzbourg; Saint Gothard; Oberwiesenthal in Saxony; Sommerleiten near Bareuth). The emeralds of Sabara, in Upper Egypt, again found by the intrepid traveller Cailliaud, and those of Salzbourg, are contained in the mass of mica-slate itself, as are, in both continents, garnets, staurotide, (St. Gothard; Sierra Nevada de Merida), and the syenite, (Shetland Isles, Maniquarez, at the north of Cumana). The emeralds of Muzo, in New Grenada, appeared to me to form a bed in a hornblende-slate subordinate to mica-slate. If we consider formations only with relation to their volume and mass, we must admit that mica-slate appears as important in the chains of the mountains of Europe, as transition porphyries and trachytes in Mexico, and in the Andes of Quito and Peru. The most considerable continued masses of mica-slate which I saw in equinoxial America, are those of the Cordillera of the shore of Venezuela, where granite-gneiss predominates, from Cape Cudera as far as Punta Tucacas (at the west of Portocabello), while towards the east the same Cordillera is composed of mica-slate, and even of a garnetiferous mica-slate, in the mountains of Macanao, the Isle of Marguerite, and in all the peninsula of Araya. At the west of of Chupa-ripari, this latter rock contains small beds of quartz, with syenite and rutile. Near Caracas granular limestone forms beds, not in the mica-slate, but in the gneiss; in the mountains of Tuy, on the contrary, there is mica-slate passing (as in the valley of Capaya) to talc-slate, containing beds of primitive limestone, and small beds of zeichenschiefer (graphic ampelite). At the south of the Oroonoko, in the mountain-group of Parime, in an extent of one hundred and eighty leagues, I saw no real mica-slate lying on granite-gneiss. The latter formation alone seems to cover that vast country, but there gneiss sometimes passes to mica-slate; at sunrise it gives a splendour to the flanks of several lofty mountains (peak of Calitamini, Cerro Ucucuamo, between the sources of of the Essequibo and Rio Branco), and has thence contributed to the mythus of Dorado and the riches of Spanish Guyana. In the Cordilleras of the Andes, the independent formation of mica-slate appeared to me less rare in the north than at the south of the equator. At Nevado de Quindiu (New Grenada) it attains a thickness of more than six hundred toises. Advancing from thence by Quito and Loxa, towards the Andes of Peru, we saw mica-slate appearing from beneath the trachytes and transition porphyries of Popayan (at the south of the volcanoes of Sotara and Purace); further on, that rock remains visible on different points from the Alto del Roble (the ridge that divides the waters between the Pacific Ocean and the Sea of the Antilles) as far as the valley of Quilquasé; it is again hid at intervals beneath trachytic porphyries, with a base of phonolite, and re-appears several times; for instance, between Almageur and the Rio Yacanacatu, between Voisaco and the volcano of Pasto, between Ganace and the volcano of Tunguragua, between Guamote and Tiscan near Alansi (where the mica-slate contains an immense bed of quartz containing sulphur, and another bed (?) of primitive gypsum), between Guasunto and Popallacta, between the Cañar and Burgay, at the southern part of the trachytic group of Assuay, finally, between Loxa and Gonzanama. Near the latter place, in the ravine of Vinayacu, a bed of lamellar graphite occurs in a mica-slate that is certainly primitive. In descending from Loxa by the Paramo of Yamoca, towards the Amazon, between the 4° and 5° of south latitude, in the valley of Pomahuaca, granite of the second formation is covered by mica-slate; but in general, in that part of the Cordilleras, it is not mica-slate, but primitive syenite and clay-slate that have assumed a great developement, whenever the soil is not covered with porphyries and trachytes. In New Spain, mica-slate abounds (gold mines of Rio San Antonio) in the province of Oaxaca; but farther north (16°—18° N. lat.), on the eastern declivity of the Cordilleras, between Acapulco and Sumpango, the granite is not even covered by gneiss, but immediately by alpine limestone (Alto del Peregrino) and transition porphyries (the Moxonera Acaguisotta). A mica-slate, destitute of garnets and passing sometimes to clay-slate, occurs, however, in the rich mines of Tehuilotepec and Tasco (between Chilpansingo and Mexico), beneath alpine limestone. Veins of red silver penetrate from one of these rocks into the other, notwithstanding the great difference that must be admitted in the ages of the rocks. I know no example in the Andes of a bed of porphyry in mica-slate, or a passage of the latter rock to a porphyritic rock; a passage which, according to the important observation of M. de Buch, takes place in the Alps of Splugen, between the village of that name and the valley of Schams. The primitive formations in which mica-slate abounds, supply mineralogists with the greatest variety of crystallised substances. These rocks, so abundant in potash, rival in this respect transition mandelstein (amygdaloids), and several volcanic rocks. We very rarely observe in nature an equal developement of the three formations of gneiss, mica-slate, and clay-slate; and when that developement has taken place, it is rather in mountains of small elevation, and where they sink towards the plains, than in the lofty chains of the Andes, the Alps, the Pyrenees, and Norway. No where, perhaps, is the total suppression of micaceous or schistose formations more frequent than in the Cordilleras of Mexico and South America. We there see the series of primitive rocks stop abruptly, either at gneiss-granite, at a syenite which I believe to be primitive, or at gneiss-mica-slate. This phenomenon occurs even where (Cordillera of Parime) there is an absence of trachytes, and every other volcanic phenomenon.

GRANITE POSTERIOR TO MICA-SLATE, AND ANTERIOR TO CLAY-SLATE.

§ 12. A granite of new formation, reposing on mica-slate, to which it belongs geognostically (St. Gothard in the Alps, Reichenstein in Silesia). It is often stratified (Högholm in Norway, according to M. de Buch; Matfreidersdorf and Striegau in Silesia, according to M. Schulze), contains garnets and hornblende, and passes to a syenitic rock with large grains; the quartz is remarkable for its great transparency, and the feldspar for the size of its crystals. The granite is sometimes steatitic, and indicates the return of slate rocks to granular and crystallised rocks. The granite of Mittlewald, north of Boixeu (passage of the Alps of Brenner), reposes on a primitive syenite, which alternates several times with mica-slate. The granite with topazes of Schneckenstein in Saxony, long considered as a particular formation (topazfels), is probably only a mass that traverses the mica-slate. I admit the existence of a formation of granite, analogous to that of St. Gothard (that is, posterior to mica-slate), in the Andes of Baraguan, Quindiu, and Herveo, where several modern granites appear on the crest of the Cordilleras, supporting peaks of trachytes. Is it to this formation that the granite of Krieglach in Styria belongs, in which the lazulite (blauspath) replaces common feldspar, and the remarkable rock of the Carnatic, the knowledge of which we owe to the Count de Bournon? This latter is composed of indianite, feldspar, and corundum (with garnets, epidote, and fibrolite).

GNEISS POSTERIOR TO MICA-SLATE.

§ 13. A small formation of garnetiferous gneiss, observed by M. de Buch. It covers mica-slate, (Bergen, Classness and Klowen, in Norway), and contains subordinate beds of granular limestone, and even of mica slate. This formation is also found in the Pyrenees.

GREENSTONE-SLATE, OR GRUNSTEIN-SCHIEFER.

§ 14. Greenstone-slate (diabase schistoide, grün-stein-schiefer,) is placed between gneiss and primitive clay-slate, (Siebenlehn, Roseathal), or between mica-slate and primitive clay-slate (Gersdorf and Rosswein, in Saxony); it contains very ancient argentiferous veins. Greenstone-slate occurs also as a bed subordinate to mica-slate. It is a formation of compact feldspar; and its independence appears to me doubtful.

IV. PRIMITIVE CLAY-SLATE.

§ 15. Primitive schist (schiste argilleux, phyllade, urthon-schiefer,) is less carburetted, and generally of a colour less dark than the transition clay-slate. Where it passes to mica-slate, the mica is in large scales, while mica in small scales characterises transition clay-slate; *Subordinate beds;* bluish granular limestone; porphyry; chlorite-slate with disseminated garnets and sphene; mica-slate (Klein-Kielvig in Norway); greenstone, but more rarely than in the transition clay-slate; greenstone-slate; quartz with epidote; a mixture of diallage and feldspar; beds subordinate in primitive clay-slate are less frequent than those in mica-slate, a rock in which the heterogeneous beds, and the abundance and variety of crystallised substances have attained their *maximum*, in passing from primitive granite to the transition rocks. When we consider, in a general point of view, the difference between primitive and transition clay-slate, we can indicate in the former several very important negative characters; such as the absence of nodules or subordinate beds of compact limestone, the absence of chiastolite disseminated in the mass, layers of clay-slate shining strongly charged with carbon, finally, the absence of numerous beds of greenstone (in balls), of aluminous and graphic ampelite (alum-slate, and drawing-slate, alaun-und zeichen-schiefer), lydian stone, and siliceous schist; but we must not forget that those general characters admit of partial exceptions, at which the experienced geognost is the less surprised, as transition clay-slate often immediately succeeds, according to the relative age of formations, to primitive clay-slate. We find in the latter, chiastolite, at the summit of the Pyrenees, and near Kielvig in Norway. M. de Raumer saw together in Silesia (Rohrsdorf, Nieder-Runzendorf), subordinate beds of porphyry with a base of feldspar, gneiss, mica-slate, granular limestone, ampelite, and lydian stone. In equinoxial America (littoral chain of Venezuela, isthmus of Araya, Cerro de Chuparipara), I observed together in a clay-slate which passes to primitive and cyanitiferous mica-slate on which it reposes, beds of rutile and shining drawing-slate traversed by small veins of native alum. It is sometimes very difficult to observe with precision where the primitive clay-slate ceases, and where the transition slate begins. The blackish-blue slates of Piedras Azules (between Villa de Cura and Parapara) at the ancient northern shore of the Llanos, or steppes of Venezuela, those of Guanaxuato, at Mexico, of which the lower strata pass to talc-slate and chlorite-slate (talc and chloritschiefer), while the upper strata are charged with carbon, and contain beds of serpentine syenite, are found on the limit of the two contiguous formations.

There is no doubt that in both continents the greatest mass of schists are those of transition; but in America, particularly in the equinoxial region, this difference is less striking than the actual scarcity of all clay-slate compared with gneiss-mica-slate. Clay-slate appears to be entirely wanting in the Cordillera of Parime, across which the

Oroonoko has worked its way ; in the Andes, as well as in the Pyrenees, it occupies a space of but small extent. I found it at the north of the equator, supporting the secondary formations of the table-land of Santa-Fé de Bogota, between Villeta and Mave; and at the south of the equator, placed on the mica-slates of Condorasto, and serving as a basis to the transition porphyries of the Alto de Pilches, between San Luis and Pomallacta (Andes of Quito); beneath the alpine limestone of Hualgayoc basseting out at a height of 2000 toises, in the Paramo of Yannaquanga (ridge of the Andes of Peru); and reposing immediately on ancient granite, between the Indian villages of San Diego and Cascas (western declivity of the Andes of Peru). I do not know if the clay-slate covering a syenite that belongs to granite, on the banks of the lake of Hacatacumba, and at Paramo de Yamoca (eastern declivity of the Andes of Peru, province of Jaen de Bracamoros,) be really a primitive formation. The insensible passages which we sometimes observe between granite, gneiss, mica-slate, and clay-slate, and which have their analogies in the passage of the syenites and serpentines to transition greenstone, has led many geognosts to believe that those four formations are only one. We see, in fact, a vast extent of country, in which gneiss constantly oscillates between granite and mica-slate; and mica-slate between gneiss and clay-slate; but this phenomenon is by no means general. We must distinguish in the two hemispheres, 1°, the districts where these insensible passages, or oscillations between neighbouring rocks, take place frequently, and in an irregular manner: 2°, the districts where distinct strata of granite and gneiss, of gneiss and mica-slate, alternate, and constitute complex formations of granite and gneiss, and of gneiss and mica-slate: 3°, the places where simple formations of granite, gneiss, mica-slate, and clay-slate are superposed without alternating (with or without a passage at the point of mutual contact). The latter case does not exclude, in the gneiss for instance, the beds of granite which resemble the lower rocks, nor the beds of mica-slate which indicate our approach to the rocks which are superposed:

We place after the clay-slate four parallel formations.

QUARTZ-ROCK.	PRIMITIVE PORPHYRY?
GRANITE-GNEISS, POSTERIOR TO CLAY-SLATE.	PRIMITIVE EUPHOTIDE.

The first of these formations is little known in Europe; the third appears doubtful as an independent formation.

QUARTZ-ROCK (WITH MASSES OF SPECULAR IRON, FER OLIGISTE METALLOÏDE).

§ 16. This is the great formation that contains the itacolumite, or elastic chloritous quartz (gelinkquarz, biegsamer sandstein, chloritquarz) of M. d'Eschwege, and the beds of micaceous and specular iron. On the south of the equator we find, in the mountains of Brazil, and in the Cordilleras of the Andes, masses of quartz, sometimes quite pure, sometimes mixed with talc and chlorite, and which, from the enormous thickness of their beds, and the extent they occupy, merit the attention of geognosts. Those rocks of quartz appeared to me to consist of several formations of very different relative antiquity. In South America, some are connected with a clay-slate decidedly primitive; others, much more difficult to understand in their relations of superposition, are placed between transition porphyry and alpine limestone, and sometimes replace the red sandstone. We shall here speak only of the former, separating the formations of which the position is well known, from those which are more uncertain. On the table-land of Minas-Geraes, near to Villa-Rica (according to the excellent observations of M. d'Eschwege, director-general of the mines of Brazil), a mica-slate that contains beds of granular limestone is covered by primitive clay-slate. On this latter rock reposes, in conformable stratification, the chloritous quartz (chloritquartz), which constitutes the mass of the Peak of Itaculumi, 1000 toises above the level of the sea. This formation of quartz contains alternating beds; 1°, of auriferous quartz, white, greenish, or striped, mixed with talc-chlorite, and exhibiting strata of flexible quartz, which have hitherto been improperly considered as hyalomicte (greisen), or as beds of quartz in mica-slate; 2°, chlorite-slate; 3°, auriferous quartz, mixed with tourmaline (schorl-schiefer of Freiesleben); 4°, specular iron, mixed with auriferous quartz (goldhaltiger eisenglimmer-schiefer). The beds of chloritous quartz are sometimes 1000 feet thick. The whole of this formation is covered by a ferruginous breccia, extremely auriferous. M. d'Eschwege thinks that it is to the destruction of the beds we have just named, and which are geognostically connected together, that the soil which is worked by means of washing, should be attributed, containing gold, platina, palladium and diamonds (Corrego das Lagens), gold and diamonds (Tejuco), platina and diamonds (Rio Abacte). The decomposed chlorite-slate, from which the topaz and the euclase of Brazil are procured, belongs to this formation. Sometimes in the mountains of Minas-Geraes, the quartz-rock is of a more simple structure. Instead of being composed of alternate beds, it consists but of a simple mass of quartz, with dense or granular specular iron (dichter eisenglanz; fer oligiste not lamellar nor micaceous). This mass is sometimes 1800 feet thick, and contains no disseminated gold. It is placed on primitive clay-slate that immediately covers gneiss. It may be said that it is this but little known formation of itacolumite quartz, which has furnished, by its decomposition, (into the alluvial soil which has proceeded from it) in the years 1756-1764, nearly thirty millions of franks in gold annually. It immediately succeeds clay-slate; but according to the observations hitherto made, it would be difficult to class it with the novaculite or whet-slate (cos, wez-schiefer), which is greenish-grey, smoke-grey, mixed with much alumine, and forming a subordinate bed in clay-slate. The itacolumite quartz, by an oryctognostic affinity which exists between talc and chlorite, is allied to talc-slate (talk-schiefer), which abounds, in every other country, with minerals well crystallised, and which, by the suppression of the plates of talc, is sometimes only pure quartz; the talc-schist forms, therefore, in the two continents, beds subordinate to clay-slate, and to primitive mica-slate. I found a formation analogous to that of Minas-Geraes, but destitute of specular iron, at the height of 1600 toises above the level of the sea, in the savannahs of Tiocaxas (at the south of Chimborazo, between Guamote and San-Luis), and at the east of the Paramo de Yamoca, near Hecatacumba (Andes de Quito). Enormous masses of quartz are there mixed with some plates of mica, and superposed to primitive clay-slate. The independence of the primitive quartz formations which we here indicate, will be better established when we shall find them superposed immediately, not always on the same rock (on clay-slate), but on different rocks more ancient; for instance, on mica-slate, gneiss, and granite. In this independence of position, the quartz-rock of Contumaza, which I believe to be secondary, is observed; it first covers porphyry, then (near Cascas), the same granite that forms the coast of the South Sea in the Lower Peru. A very important observation made by M. de Buch in the north of the Scandinavian peninsula, appears to justify the position we assign, among primitive rocks, to the quartz-rock of the southern hemisphere. That indefatigable traveller noticed, that in the northern region of the ancient world, primitive clay-slate is sometimes replaced by a quartz-rock coloured by iron. Quartz-rock and clay-slate are consequently in Norway parallel rocks, or geognostic equivalents. It is very remarkable to see sulphur, gold, mercury, and specular iron ore, connected in South America with those enormous masses of silex. Whatever interest the precious metals may excite, it cannot be denied that the abundance of sulphur in primitive formations is, with respect to the study of volcanoes, and those rocks through which the subterraneous fires pierce their way, a far more important phenomenon than the abundance of gold. A little to the south of the elevated savannahs of Tiocaxas and Guamote (Cordilleras of Quito), where we just noticed the formation, perhaps independent, of quartz superposed to clay-slate, I examined the celebrated mountain of sulphur of Tiscan, which is a bed of quartz (direction N. 18° E.; inclination 70—80″ at N. W.; thickness of the bed 200 toises; height above the level of the sea, 1250 toises) in mica-slate. In Brazil, the formation of chlorite-quartz (itacolumite), superposed on primitive clay-slate, contains not only gold, but sulphur also; slabs of this rock, strongly heated, burn with a blue flame. A clay-slate of the same age as that on which chloritous quartz is superposed contains (Serro de Frio, near S. Antonio

Pereira) a bed of primitive limestone containing masses of native sulphur. Gold and sulphur are found also (Andes of Caxamarca, at Peru, between Curimayo and Alto del Tual), on the limit of transition porphyries and alpine limestone, in considerable masses of quartz which are parallel to the red sandstone. To these quartz-rocks, or rather to still newer formations, belong the great deposit (quarzflotz) of sulphuret of mercury of Grancavelica, while the mercury of Cuença (southern part of the kingdom of Quito), as well as that of the duchy of Deuxponts, belongs to the red sandstone. These notions will serve to throw some light on the great beds of quartz which M. d'Eschwege and myself observed in the southern hemisphere, and which can scarcely be called quartzose sandstone. These rocks appear to pass, like limestone, through different primitive, intermediary, and secondary formations. Several celebrated geognosts have already attempted to introduce quartz-rocks as independent formations, in the general type. The *quarzgebirge* of Werner is primitive, and reposes on gneiss (Frauenstein, Oberschönau, in Saxony), by which, perhaps, it was formerly covered. Beds that belong essentially to one formation are sometimes found at the upper and lower limit of that formation; (for example, bituminous slate below zechstein, or alpine limestone; gypsum above zechstein; siliceous schist, lydian stone, or ampelite, above transition clay-slate, and in that rock). The small masses of primitive quartz, observed on the ridge of the mountains of Europe, cannot be compared in thickness nor extent to the rocks of primitive quartz of the Andes or Brazil. The granular quartz-rock (with feldspar), described by Mr. Jameson, in the Hebrides, the quartz and chlorite-rocks, anterior to grauwacke and connected with red sandstone (primary red sandstone of Dr. Maculloch), afford some traits of geognostic analogy with the quartz masses of equinoxial America; but they are much more mixed (less simple in their structure), and may, according to the interesting discussions of M. Boué, very probably belong to ancient transition rocks. The trapsandstein, or secondary quartz-rock of some German geognosts, surround basalts, and are, no doubt, of an origin much more recent than the formation of quartz in large masses (extremely pure, without mixture, and not aggregated), which, placed between transition porphyry and alpine limestone, attain, according to my observations at the western declivity of the Andes of Peru (Contumaza, Namas), the enormous thickness of 6000 feet.

GRANITE AND GNEISS, POSTERIOR TO CLAY-SLATE.

§ 17. This is a formation of small grained granite, passing sometimes to a garnetiferous gneiss, with which it alternates. This interesting formation reposes, (Kielvig, at the northern extremity of Norway, and the Shetland islands), according to M. de Buch, on primitive clay-slate. It contains hornblende and diallage, and thus manifests its affinity with one of the following formations. The formations of granite might be designated (§§ 4. 7. 12. and 17.) by the names of granite of whitestone (weistein), of gneiss, of mica-slate, and of clay-slate; but such denominations would lead to the supposition that these small formations are necessarily situated in whitestone (weistein), in gneiss, in mica-slate, and in clay-slate; whereas they are found simply superposed on these rocks, on which they seem to depend. The presence of tin, magnetic iron (?), hornblende, diallage, garnet, talc and chlorite replacing mica, as well as the tendency to pass to pegmatite (schriftgranit) characterises the newer granites.

PRIMITIVE PORPHYRY?

§ 18. Does a primitive and independent formation of porphyry exist? There can be no question respecting the porphyries which occur as subordinate beds in other primitive rocks (§§ 5. and 15.), nor with respect to the gneiss and mica-slate of the high Alps, which become granular, and assume by the insulation of feldspar crystals, a pophyroidal aspect. I hesitate to place the porphyries of Saxony and Silesia (duchy of Schwiednitz) among primitive rocks, although the former immediately cover gneiss (between Freiberg and Tharandt. They are sometimes traversed by veins of tin (Altenberg) and ores of silver (Grund). The porphyries of Silesia contain disseminated hornblende (Friedland); and have hitherto been considered as more ancient than primitive clay-slate. It is certain that the porphyries of Saxony are partly transition porphyries, and partly belong to the red sandstone. In the Cordilleras of the Andes of Peru, Quito, New Grenada, and Mexico, among that innumerable variety of porphyritic rocks, of which the masses attain from 2500 to 3000 toises in thickness, I did not see a single porphyry that appeared to me decidedly primitive. The most ancient formation which I observed occurs in the deep valley of Magdalena, (between Guambo and Truxillo, in Peru): it is a porphyry with a clay basis a little decomposed, with common not vitreous feldspar, without hornblende, but also without quartz. This formation, which appears distinct from all the transition porphyries and trachytes of Quito, and the ridge of the Andes of Peru, appears at the height of 600 toises above the level of the sea; it is placed immediately upon granite, and is covered, at the western declivity of the Andes by a rock of secondary quartz, and (probably) at the eastern declivity by red sandstone.

V. PRIMITIVE EUPHOTIDE POSTERIOR TO CLAY-SLATE.

§ 19. This formation is placed at the limit of primitive and transition formations. It is the gabbro of M. de Buch; the euphotide of M. Haüy; the schillerfels of M. de Raumer; and the ophiolite of M. Brongniart. This rock was formerly designated by the names of serpentinite, serpentinous granite, granite of diallage granitone, granito di gabbro, granito dell' impruneta, serpentinartiger ur-grünstein. We shall here describe it such as it was first limited by M. de Buch. It is found superposed (cape north of the isle Mageroe, in Norway) in a primitive slate, which passes towards the upper part to euphotide, and towards the lower to mica-slate. The euphotide of Val Sesia covers also, immediately according to M. Beudant, primitive mica-slate. It may be said that in general the euphotide or gabbro is a mixture of diallage (smaragdite), jade (saussurite, tenacious feldspar), and lamellar feldspar. Sometimes (Bergen, in Norway) jade is entirely wanting; but in the Verde di Corsica (Stazzona, at the north of the Corte and S. Pietro di Rostino in the island of Corsica) euphotide is only a mixture of jade approaching to compact feldspar and of green diallage without lamellar feldspar. Although, according to the interesting observations of M. Haüy in his *Tableau Comparatif*, the diallage metalloid (schillerspath), whch is green with a silky lustre, and the grey diallage pass progressively (rocks of Masinet near Turin) into each other; these substances may, however, be distinguished by the geognostic characters which they most frequently exhibit in great masses. The euphotide with grey diallage is much more frequent (a little more ancient?) than the euphotide with green diallage. Serpentine is almost always intimately connected in position with euphotide, of which it seems to be only a variety, with small grains of homogeneous appearance. This connection is also manifested in Hungary (Dobschau), where M. Beudant found green and slaty euphotide immemediately superposed on primitive mica-slate. Soda, according to the labours of Theodore de Saussure and Klaproth, occurs among primitive rocks in the compact feldspar of whitestone and greenstone slate, in the jade of the euphotide, and in the lazulite of Baldakschan. The latter substance appears to belong to a bed of primitive limestone interposed in granite-gneiss. Subordinate beds to euphotide: serpentine with asbestos, and diallage metalloid; serpentine accompanied by chrysoprase, opal, and calcedony, (Kosemitz, in Silesia); compact greyish limestone passing to fine-grained limestone, (Alten, in Norway). This limestone connects the euphotide of Scandinavia, which is the last member of the primitive formations, with very ancient intermediary rocks. Euphotide being often not covered, and the superposition of a rock on another very ancient rock throwing no light on the epocha of its formation, doubts remain of the relative age of many euphotides. M. de Buch saw that of the Upper Valais (Saas, Mont-More) placed above mica-slate; and that of Sestri at the north of the gulph of Spezzia, below clay-slate (of transition?), in the Zaragua: M. de Raumer, in his excellent work on Lower Silesia, places the schillerfels (rock of schiller spar) of Zobtenberg among the primitive formations. M. Keferstein places among them the euphotide of the Hartz (between Neustadt and Oderkrug), which contains disseminated ferriferous titanium (nigrine). I think, also, that the serpentines of Heideberg, near Zell, and those found between Wurlitz and Kotzau, where they contain pyroxene-diopside, are very ancient. All the serpentines of the mountains of Bareuth appeared to me to be closely connected with hornblende-slate (hornblend-schiefer) and chlorite-slate (chlorit-schiefer). They possess very remarkable magnetic properties, which I made known in 1796, and which have since been the object of the more accurate researches of M. M. Goldfuss, Bischof, and Schneider. In taking a general view of the euphotides of both continents, we cannot but admit several formations of very different ages. The euphotides which I observed at the Isle of Cuba, Guanaxuato, Mexico, and the entrance of the Llanos of Venezuela, are connected either with syenite, or black limestone, and appeared to me to be decidedly transition euphotides, the same as the euphotide (serpentine stratified in their beds; direct. N. 52° E.; incl. 70° at the N. W.; thickness 10 toises) of the summit of the Bochetta of Genoa, which I observed in 1795 and 1805, and which is imbedded in transition clay-slate that alternates with black limestone. The euphotides of Spezzia, Prato, and the whole of the Siennese, which MM. de Buch and Brocchi consider as of primitive or very ancient transition formation, appear to M. Brongniart, by whom they have recently been examined with great care, to belong to secondary formations, or at least to the most recent transition formations. The celebrated geognosts whom I have just named agree on the immediate position of these euphotides of Italy, that is, on the oryctognostic determination of the rocks which are found below and above the euphotide; but they differ respecting the date which ought to be assigned to the rocks in contact with it. Thus, in geography, we sometimes know with precision the position of an island with respect to the neighbouring islands; while the absolute longitude of the whole group, and its proximity to the old or new continent, may remain uncertain.

TRANSITION FORMATIONS.

The Transition formation, according to M. Werner, consists of rocks which exhibit in their composition a great analogy with those of the primitive formation, but which alternate with brecciated or arenaceous rocks (*aggregated clastique* rocks of transport). Some remains of organised bodies (impressions of reeds, palms, arborescent ferns, madrepores, pentacrinites, orthoceratites, trilobites, hysterolites, &c.) appear chiefly, I will not say in the upper rocks, or the least ancient of that order, but in general in the non-feldspathic rocks, the mass of which does not exhibit a very crystalline aspect.

The excellent observations of MM. de Buch and Brochant have contributed most to extend the limits of the transition formations. Those limits are more easily fixed towards the upper part, where the secondary formations begin, than towards the lower, where the primitive formations end. I have elsewhere observed in what manner, through the anthracitous mica-slate and green clay-slate, transition rocks are connected with primitive rocks; how, through the porphyries with glassy feldspar, they are allied to volcanic rocks; and by fine-grained grauwacke and porphyries abounding in crystals of quartz, to the red sandstone and secondary porphyries. In regions the most distant from each other, analogous rocks, such as talcose clay-slate, with foliæ much contorted, and carboniferous, containing ampelite (alaun-schiefer) and lydian stone; black limestone alternating with mica-slate, grauwacke, and porphyries; and syenites mixed with titaneous iron; are placed between primitive rocks (that is, rocks destitute of all traces of organised bodies, and of arenaceous masses,) and the great coal formation; but the succession of homonymous transition rocks varies even where they all seem to be equally developed. The greatest number of those formations are composed of two or three alternating rocks (compact black limestone, greenstone, and clay-slate; grauwacke, and porphyry; granular limestone, grauwacke, and mica-slate, with anthracite); and as the partial members of groups or formations of so complicated a structure pass from one group to another, some excellent observers, MM. de Raumer, d'Engelhardt, and Bonnard, have been so much struck by this phenomenon of connection and alternation, that they recognise in the whole class only one great family of rocks. If we examine the transition formations according to their structure and their oryctognostic composition, we distinguish five very strongly marked associations; the schistose rocks; the porphyritic rocks (feldspathic or syenitic); granular and compact limestone, with anhydrous gypsum, and rock-salt;

euphotides and the aggregated rocks (grauwacke, and calcareous breccias). On some points of the globe, one only of those groups or associations of crystallised and non-crystallised rocks has received so extraordinary a development that the other groups appear almost entirely suppressed. Thus in the Cordilleras of Mexico and Quito, as well as in Hungary and several parts of Norway, the porphyries and transition syenites predominate; in the Tarentaise, granular and talcose limestone; in some parts of the Alps, and at Bochetta, black limestone, almost compact, or small-grained; finally, in the Hartz, and on the banks of the Rhine, grauwacke and transition clay-slate; but this thickness and extent which the mineral masses attain to ought not to determine the geognost when he discusses the relative age of partial formations. A very great variety of position is observed not only in the small formations, but the great homonymous formations also, when much developed, can scarcely be considered as contemporary; that is, they do not hold the same position with respect to the other terms of the series of the intermediate rocks. The porphyries of Guanaxuato, for instance, are situated on a steatitic and carboniferous clay-slate, and those of Hungary rest on a talcose mica-slate of transition, containing beds of dark-grey limestone. The porphyries of the Andes of Quito, (and the British Islands?) repose immediately on primitive rocks, and are consequently anterior to every limestone rock that contains vestiges of organised bodies: on the contrary, the porphyries and zircon-syenites of Norway, and probably also the porphyries of Caucasus, so well examined by MM. d'Engelhardt and Parrot, succeed, according to the age of their formation, to limestone filled with orthoceratites. The most considerable masses of grauwacke (alternating with grauwacke-slate) are no doubt developed amongst the most ancient transition slates; but we also find very large beds of grauwacke of a much more recent origin. In general, the five groups of rocks which we have just distinguished according to the relations of composition, or the oryctognostic characters, do not preserve every where the same place in the series of intermediate formations; they are not found distinctly separated in nature as in a oryctognostic classification. Clay-slate and black limestone, clay-slate and porphyry, clay-slate and grauwacke, porphyry and syenite, granular limestone, anthracitous mica-slate, are observed to form geognostic associations in countries the most remote from each other. The constancy of these binary or ternary associations characterises the transition formations, much more than the analogy which the succession of homonymous rocks presents in every group.

In a discussion on the primitive series of rocks where the formations are more simple, better defined, and less subject to frequent alternations, I have attempted to enumerate separately, the granites which succeed to gneiss, and the gneiss which succeeds mica-slate. There are primitive granites and gneiss of different ages, as in the transition formations grauwacke or black limestone occur, similar in composition, but very distant from each other in their relative antiquity. If in these latter formations, the geognost does not attempt to name separately the different beds of grauwacke or of limestone, it is because these beds separately have no value as terms of the series of intermediate rocks, but only inasmuch as they make a part of certain groups. But these groups, those constant associations of clay-slate, greenstone, and grauwacke, of steatitic limestone and grauwacke, porphyry and grauwacke, &c., are the true terms of the series. It thence results, that, according to the principles which we follow in the arrangement of formations, we ought to enumerate separately, not insulated masses of limestone, of grauwacke, and of porphyry, which are mixed together or with other rocks, but entire and well characterised groups; those, for instance, in which grauwacke and clay-slate, porphyry and syenite, predominate. Some among these last are posterior, and others are anterior to the rocks that contain vestiges of organised beings. In the primitive formations the terms of the series are generally simple; in transition formations they are all complex; and from this complicated nature arises the difficulty of studying, step by step, an edifice of which we can with difficulty comprehend the order amidst a confusion of similar parts. To justify the order which I assign to the different transition formations, I shall begin by presenting in the following table the succession of formations (beginning with the most ancient) which have been observed in several countries, and carefully examined. I shall employ only the oreographic description of geognosts who have been accustomed to follow the same principles in the denomination of rocks.

1. Andes of Quito and Peru.

Transition porphyries, not metalliferous, covering immediately the primitive rocks (granite, clay-slate).

Greenstone in balls (kugelgestein).

Black limestone superposed on porphyry.

I saw no grauwacke; it is replaced in the Andes of Quito and Peru, at the south of the equator, by the great porphyry formation.

2. Mountains of Venezuela.

Green steatitic transition slate, covering primitive gneiss-mica-slate.

Black limestone.

Serpentine and greenstone, covered by amygdaloid, with pyroxene.

This is the succession of rocks which I observed at the northern bank of the Llanos of Calabozo.

3. Mountains of Mexico.

Transition clay-slate, with carbon, and containing, also beds of syenite and serpentine. The lower beds pass to talc-slate, and repose on primitive rocks.

Syenite alternating with greenstone.

Transition porphyry, metalliferous and placed immediately on transition clay-slate. The upper beds pass into phonolite.

Such is the series of rocks of Guanaxuato. In the road from Mexico to Acapulco, I saw the transition porphyries resting immediately on primitive granite. Near Tutonilco these porphyries are covered by secondary rocks, such as alpine limestone, sandstone, and gypsum, mixed with clay. I shall not decide on the age of the transition limestone of the mines of the Doctor, and Zimapan, nor on that of the porphyries of Guanaxuato and Pachuca; but, according to MM. Sonneschmidt and Valencia, in the rich mines of Zacatecas, nearly as at Guanaxuato, syenite and transition clay-slate (with greenstone and lydian stone), grauwacke, and non-metalliferous porphyry, follow in succession from below upwards.

4. Hungary.

Transition mica-slate, with beds of black limestone, resting on primitive rocks.

Transition porphyries and syenites. Subordinate beds: transition mica-slates; white granular limestone with serpentine; masses of greenstone. These porphyries, like the greatest part of those in the Andes, are immediately covered by white and black syenitic trachytes. (*Observations of M. Beudant.*)

5. Tarentaise.

A single formation, reposing immediately on primitive rocks, contains steatitic granular limestone, mica-slate with gneiss, and anthracitous grauwacke. These different rocks alternate several times, and exhibit subordinate beds of serpentine, greenstone, compact quartz, and transition gypsum. (*Observations of M. Brochant de Villiers*).

6. Switzerland.

In the passage of the Alps, from Chiavenna to Glaris, according to M. de Buch: —

Transition clay-slate, with beds of grey limestone, reposing on clay-slate, and primitive mica-slate.

Serpentine with garnets.

Black limestone.

Grauwacke.

Clay-slate, alternating with black limestone.

Clay-slate, with impressions of fish (almost secondary).

In the vicinity of Bex, according to M. de Charpentier.

Grauwacke, superposed or (primitive?) gneiss.

Black limestone, containing belemnites, and alternating with transition clay-slate.

Argillaceous limestone of transition with ammonites, subordinate beds of grauwacke, anhydrous gypsum, and rock-salt.

M. de Buch, according to geognostic observations made before the year 1804, assigned to the transition formations in the west of Switzerland, considered in a general point of view, and in passing from the lower to the upper rocks, the following order: —

Transition clay-slate, black limestone, saliferous muriacite, and gypsum; grauwacke, black limestone, clay-slate, with impressions of fish.

7. Germany.

System of position in Saxony, between Freiberg, Maxen, and Meissen, according to MM. de Raumer and Bonnard.

Clay-slate, with ampelite and lydian stone, alternating at the same time with grauwacke, greenstone, porphyry, and limestone. This formation reposes on primitive gneiss.

Syenite and porphyry. In this formation, which abounds also in Thuringerwald, according to the excellent description of M. Heim, transition gneiss and granite are found subordinate.

The Hartz and Western Germany, (between the Rhine and the Lahn,) contain a great formation of clay-slate, in which, as if by interior development, masses of grauwacke, and grauwacke-slate, limestone (often of a light colour), greenstone, quartz, and porphyry appear. The latter rock is here more rare, however, than in the independent formation of syenite and porphyry, which supports, in other countries, transition clay-slate.

8. Peninsula of the Cotentin and Bretagne.

Green shining steatitic (clay-slate of transition), alternating sometimes with grauwacke, with black limestone, and with quartz-rock.

Syenite and granite.

Transition clay-slate, sometimes again covering syenite. (*Observations of MM. Brongniart and Omalius d'Halloy.*)

9. British Islands.

Transition-syenite and porphyry reposing on primitive rocks, (Snowdon, the Grampians, Ben-Nevis).

Transition clay-slate, with trilobites, containing, in the lower beds, a conglomerate of primitive rocks, similar to that of Valorsine, (Llandrindod, Killarney, summit of Snowdon).

Grauwacke, (May-hill, and North Wales).

Transition limestone, (Longhope, Dudley).

Grauwacke, old red sandstone, (Mitchel-Dean in Herefordshire).

Transition limestone, mountain-limestone (Derbyshire) covered by the great coal formation. (*Observations of Mr. Buckland*, who seems, however, to regard the syenite and part of the porphyries as primitive.)

10. Norway.

Position of the rocks near Christiania, according to the observations of M. de Buch.

Transition clay-slate, alternating with black limestone with orthoceratites, and reposing on primitive gneiss, grauwacke and kiesel-schiefer.

Porphyry with crystals of quartz, containing a bed of porous greenstone, with pyroxene, zircon-sienite, and transition granite with porphyritic beds.

11. Caucasus.

Clay-slate, perhaps transition black limestone with ampelite.

Transition porphyries, alternating with clay-slate. This porphyry, often columnar with vitreous feldspar, a little quartz, and little mica, resembles, in the mountains of Kasbek (as porphyries do often on the Mexican summits) porous trachyte.

Transition gneiss, syenite, and granite in alternating beds.

Transition clay-slate, covered with fetid limestone, which appears to be secondary. (*Observations of MM. Engelhardt and Parrot.*)

We may perceive in these different types of superposition, collected in Europe, America, and Asia, at the north and south of the equator, that among the most ancient transition rocks, three great formations, that of talcose granular limestone, grauwacke with anthracite and mica-slate, that of syenite and porphyry (with crystals of hornblende, and very little quartz), and that of clay-slate, grauwacke, and black limestone, occupy nearly the same rank on different points of the globe. The micaceous limestones, and pudding-stones, with fragments of primitive rocks of Tarentaise; the porphyries and syenites of Peru; the transition clay-slate with grauwacke (Hartz, Friedrickswalde in Saxony, Aggersely in Norway, and Guanaxuato in Mexico), are perhaps of contemporary origin. In ranging the rocks as terms of a single series, we ought perhaps to have called to mind their parallelism in the following manner: II (I or III). I consider as terms of the series of transition rocks six groups, which appear to me to be well characterised by the predominating rocks, by their position, and by the extent of their masses. These groups or great formations are, I. Steatitic, granular limestone, transition mica-slate, and grauwacke with primitive fragments. II. Porphyry (not metalliferous) anterior to orthoceratite limestone, to transition clay-slate, and mica-slate. III. Clay-slate containing grauwacke, limestone, porphyry, and greenstone. IV. Porphyry and syenite (metalliferous) posterior to transition clay-slate anterior to limestones containing organic remains. V. Porphyry, syenite and zircon-granites (not metalliferous) posterior to clay-slate and limestone with orthoceratites. VI. Transition euphotide with jasper and serpentine. Almost every group is composed of alternating rocks, and several of those rocks, which may be considered as small partial formations, are common to all the groups. It is this usual occurrence, this alternation, this periodical return of the same masses, which constitute the apparent unity of the great family of transition formations. In each group, however, certain rocks predominate that give it a peculiar aspect. Such are the talcose and granular limestones in the first group; non-metalliferous porphyries, abounding in hornblende, and almost destitute of quartz, in the second; grauwacke in the third; serpentine rocks in the sixth. The fourth and fifth groups are characterised, one by metalliferous porphyries and syenites; the other by zircon-granites. But these characters are partly oryctognostic; the real basis of the division which we propose provisionally to geognosts, are the superposition and relative age observed in different parts of the globe. One part of the Mexican and Peruvian porphyries of the second and even of the fourth group, appears to have an intimate relation with trachytes, which are the most ancient among volcanic rocks.

Before I describe in detail the six great intermediary formations, I shall mention some general considerations on the transition formation, which is most frequently placed in conformable position on primitive rocks. Magnesia, oxidulated iron (the magnetic), which furnishes such striking geognostic relations with every substance in which magnesia predominates; titaneous iron; carbon, and carbonate of lime, penetrate through the greater part of transition formations. M. Beudant has made the important observation, that the syenite and porphyry of Schemnitz, Plauen, and Guanaxuato effervesce with acids, whilst the trachytes (trachytic porphyries) of Hungary do not present the same phenomenon. Saussure and M. Brochant found transition mica-slates that effervesced, (at la Tête-Noire), as well as compact quartz, (in the Tarentaise), even where those rocks are very distant from the interposed beds of steatitic granular limestone. I saw in the Cordilleras of Peru (Paramo de Yamoca), and also in the Thüringerwald-Gebirge (between Lauenstein and Grafenthal), a clay-slate which exhibited at first all the characteristics of a primitive rock, but became effervescent by degrees, and contained in its last beds dispersed nodules of darkish-grey compact limestone. The carbonate of line, at first disseminated through the whole mass, becomes gradually condensed so as to give the rock a glandular structure, and forming thin alternating strata, and interposed beds, and finally, granular or compact limestone rocks, that replace clay-slate, mica-slate, and euphotide, in the midst of which they were developed. M. Steffens, in his Treaty of Oryctognosy, has made some ingenious observations on the mutual relations of feldspar and hornblende in primitive, intermediary, and transition formations, and in the red sandstone. In the midst of the second of those formations, feldspar appears even in compact limestone. We may suppose that in passing from granite to clay-slate, through gneiss and mica-slate, this substance remains hid in the paste, which is only homogeneous in appearance; since we see transition clay-slate sometimes become porphyry, as by other internal developments, by accumulations of silica and carbon, and by the aggregation of the elements of hornblende, it becomes siliceous schist, anthracite, greenstone, and syenite. We distinguish often two sorts of feldspar in transition porphyries, the common, and the vitreous, with lengthened crystals, (Andes of Peru, valley of Mexico). The latter, which is less a mineralogical species than a particular state of common feldspar, belongs to transition-formations as well as to real trachytes. The frequent presence of hornblende, and the want of crystallised quartz, distinguish oryctognostically many transition porphyries from those of primitive formations. The latter are perhaps only subordinate beds to other rocks. Hornblende, which is almost confined to interposed beds in primitive formation, is no where more abundant than in the transition and trachytic formations. Among the former, greenstone and syenite, by the change of proportions in the elements of the crystalline substances, produce a kind of struggle between feldspar and hornblende. The pyroxene, which has been too exclusively thought to characterise the trachytes, basalts, and dolerites, is found in several transition porphyries of the Andes and Hungary. It is also found in the black vesicular and basaltic beds of the zircon-syenite of Norway. I thought I had observed in some transition porphyries of North America, traces of olivine, but they were no doubt only varieties of pyroxene, less dark and greenish, of which the dihedral summits can scarcely be distinguished, and the fusibility of which I could not try by the blow-pipe. Olivine belongs properly to basaltic formations; it is even doubtful if it occurs in the trachytes. The frequent tendency to crystallisation, which is observed in transition formations, amidst sedimentary and aggregated rocks, is so extraordinary a phenomenon, that some celebrated geognosts have been tempted to admit that many of those rocks which appear aggregated (under the form of breccias, or pudding-stones; clastic and arenaceous rocks; transition sandstone, or conglomerates), far from containing the remains of preexistent rocks, are only the effect of a confused but contemporary crystallisation. Masses which have in some strata been considered as distinct angular fragments, seem insensibly to dissolve at a little distance into the paste of the rock. Other masses, which resemble rolled pebbles, become nodules adhering strongly to the curved laminæ of a schist, and then lengthen and vanish by degrees. When we compare together certain granites and porphyries, limestone breccias, grauwackes, and red sandstones, we imagine that we perceive in rocks of an age so different, from certain indications of structure, the insensible passage of a contemporary formation from a crystallisation, simultaneous, but disturbed by particular attractions to a real aggregation (agglutination) of the débris of pre-existing rocks. In every zone granites occur with large grains, in which masses with small grains very micaceous are found included occasionally, and which appear, at first sight, to contain fragments of older granite. This appearance is as fallacious as that of many transition porphyries, euphotides, and limestones, which antiquaries and sculptors designate by the name of breccia, or regenerated rock. The pretended fragments, often striped and streaked (in the verde antico, and the limestones most sought after for decorating the interiors of edifices), are probably only masses which were the first consolidated in a fluid agitated strongly. The congealed waters of our rivers, and various mixtures of salts in our laboratories present analogous phenomena. The manner in which the re-united and angular fragments of grauwacke, those of calcareous pudding-stone with a granular paste and compact fragments, and those of certain red sandstones, appear sometimes to vanish and dissolve in the whole mass, are much more difficult to explain in the actual state of our knowledge. It cannot be doubted, that the frequent alternation of strata obviously aggregated, and strata almost homogeneous, or slightly nodular, as well as the passage of these masses to each other, have been determined by very accurate observations; and M. Bonnard was right in saying, in his Treatise of Formations, "that this phenomenon is one of the most incomprehensible which we meet with in the study of geognosy." Ought we to admit that when the outline of imbedded fragments disappears almost entirely, there has been but a short interval of time between the *solidification* of the fragments and that of the paste? We shall see shortly, in red sandstone, crystals of feldspar occurring in that very paste, and causing it to resemble the porphyry of the red sandstone. (Steffens, *Geognostisch Gealog. Aufs.*, p. 15, 16. 23. 31. Freiesleben, *Kupfersch*, tom. iv. p. 115.)

I. GRANULAR TALCOSE LIMESTONE, TRANSITION MICA-SLATE, AND GRAUWACKE WITH ANTHRACITE.

§ 20. The same formation, the same group contains different limestones, and schistose and fragmentary rocks, alternating with each other. This formation is not composed of those insulated rocks (like the formation of porphyry, syenite, and greenstone), but of three partial formations, three series or systems of rocks. The most complicated type of this grouping of almost contemporary rocks is developed at the south-east of the Alps, in the valley of Isere, where it has formed the object of the profound researches of M. Brochant. If almost all the terms of the series of intermediary rocks are complex, these terms or great formations do not vary the less according to the degree of that complex disposition, and according to the number and nature of the alternating masses. The *group of the Tarentaise* (the same by which we shall designate the formation § 20.) displays in its structure and composition (in its granular and talcose limestones, in its gneiss and mica-slates,) so much the appearance of a primitive formation, that we only recognise its relative age by some remains of organic bodies, and by the frequent interposition of arenaceous beds (pudding-stones, breccias, and grauwackes). During a long time, therefore, geognosts, neglecting the observation of alternation, and the unity of that complex formation, placed the pudding-stones of Valorsine among primitive rocks, and considered them merely as a local phenomenon. Later and more widely extended researches have made us acquainted with many analogous facts. These pudding-stones with primitive fragments are grauwackes that alternate with micaceous limestone, or with green clay-slate, or with transition gneiss. They are observed in the Alps (Trient au Valais), in the Tarentaise, in Ireland in the mountains of Killarney and St. David; finally, on the eastern coast of Egypt, in the valley of Cossier (Qozir). The limestone of the Tarentaise, and the Little St. Bernard, which contains disseminated crystals of feldspar, and which constitutes a species of porphyroid rock with a limestone basis, is found in analogous formations of the Alps of Carinthia. This phenomenon of the association of lime and feldspar is so much the more remarkable, as lamellar feldspar, and granular and compact limestone, appear to manifest every where else in their geognostic relations, a kind of repulsion much stronger than what is observed in some countries between hornblende and limestone. Transition mica-slates and gneiss have been long considered as belonging exclusively to the region south-west of the Alps; but they are also found in the clay-slate and porphyry formations of the Caucasus, and in the porphyry and syenite formations of Saxony and Hungary. In general, however, the formation which is the subject of this article, and which is characterised at the same time by the absence of porphyries, and the frequency of talcose and granular limestone, micaceous quartz, and anthracites, appears to have favoured the development of transition mica-slates and gneiss, more than the great formations of porphyries and syenites, or of clay-slate and grauwacke. In the two latter, on the contrary, we find in the greatest abundance, transition granites, which are crystalline, granular, without being laminar, almost destitute of mica, and geognostically belonging (even when they contain no trace of hornblende) to syenite, as transition mica-slates and gneiss belong to micaceous quartz. Syenites, whether simply forming beds in green clay-slate, or, constituting with porphyries an independent formation, announce the commencement of transition granite; the compact or slaty quartz containing scales of mica (quartz of the anthracitous limestone formation, quartz of transition clay-slate, and porphyry,) precede transition mica-slate and gneiss, which has been justly designated as porphyroid mica-slates with crystals (and nodules) of feldspar. These various modes of development of granites amidst syenitic rocks, of gneiss and mica-slate amidst quartzose rocks, lead us to understand why gneiss and mica-slate are found associated (vicinity of Meissen, in Saxony, and western declivity of the Caucasus,) much more rarely with granite of the transition formations than with the primitive formations. It may be said, that the granites of the first of these formations are only beds of syenite having no hornblende, and that the greater part of transition mica-slates present only modifications (in certain states) of a micaceous quartz in which mica becomes more abundant. These changes, however, by internal development, are not always made in the same manner. Sometimes, also, (valley of Müglitz, in Saxony,) transition granite rises immediately amidst clay-slate, and the syenites of Meissen and Prasitz pass at the same time to transition granite and gneiss.

The following is the alternating series of limestones, schists, and arenaceous rocks, which constitute the formation that we place at the head of the transition groups.

Talcose granular limestone, often veined, slaty, fetid (like the granular white marble of the isle of Thasos), mixed with grains or nodules of quartz, and containing (Sainte Foix) beds of transition serpentine. *Compact limestone,* yellowish, sometimes grey, and containing crystals of feldspar (Bonhomme, Little St. Bernard, and valley of the Tarentaise. Pudding-stones, or *conglomerate limestones,* with a granular paste and compact fragments (Tarentaise breccia of Villette). These three rocks, which form a subdivision of group § 20., alternate with each other, and with the schists of the following series. Compact transition limestones sometimes resemble the Jura limestone, and sometimes pass to a fine-grained limestone. The talcose saccharoid limestone, often white and veined, assumes the aspect of the beautiful primitive marble of Pentelicus (Cipolino), Hymettus, and Carystes in the island of Eubœa. The remains of organised bodies are generally wanting in this limestone series; but, as we shall soon see, the rocks of that series alternate with schists containing impressions of monocotyledon plants. M. Brochant has even discovered a petrifaction of a nautilus, or of an ammonite, in the calcareous pudding-stones of La Villette, between Moutiers and St. Maurice.

Transition clay-slate, either striped, and containing laminæ of interposed limestone, or unctuous mixed with fibrous talc (mine of Pesey), without any visible calcareous parts, but effervescing with acids. This clay-slate contains (Bonneval) subordinate beds of greenstone.

Compact quartz, or quartzites, either without mixture, or micaceous, and belonging to granular limestone, as well as to transition clay-slate. It is from the accumulation of mica in these masses of compact quartz that the mica-slate of this formation arises, and even gneiss; for quartz often contains a little feldspar disseminated in the mass. Mica-slates, passing to black bituminous schists with impressions of plants, (Montagny, Little St. Bernard, Landry,) are associated with anthracites, and alternate (Moutiers) with steatitic limestone and grauwackes, or pudding-stones with primitive fragments. The paste of these conglomerates, in which quartz, granite, and gneiss, are imbedded, is not always of the nature of clay-slates, as in the grauwackes of the Hartz (of the great formation, §22.); it most frequently resembles mica-slate. When the fragments become very rare in the mass, these rocks are confounded with real transition mica-slate.

In this group, composed of so many beds alternating periodically, the schistose series with anthracite appears a little newer, when we consider the great masses, than the limestone series. If, on one hand, the gypsums of the Tarentaise and l'Allée-Blanche, containing muriate of soda, sulphur, and anhydrous sulphate of lime, repose simply on transition formations, without being very visibly covered, it appears no less certain, from the interesting discussions of M. Brochant, that the gypsum of Cogne, Brigg, and St. Leonard in the Valais, are interposed in transition limestone itself. The great formations, § 20. and 25., are the only intermediary rocks in which porphyries and syenites do not appear to be developed; and they are also those in which saccharoid white limestone and masses of talc abound most. Lamellar feldspar, which penetrates into calcareous rocks (feldspathic calciphyres of M. Brongniart), appears to belong only to the formation § 20. Anthracites are common to this and the great formation of clay-slate and grauwacke, § 22.; but they are less frequent in the latter formation, where the carbon is rather disseminated in the whole mass of the clay-slates, lydian stones, and limestones, which it colours black, than concentrated in particular beds. Anthracite, as M. Briethaupt has well observed, is a more ancient formation than coal, and a more recent formation than graphite, or carburetted iron. Carbon becomes more hydrogenetted in proportion as it approaches the secondary rocks. These rocks bear the same geognostic relation to coal as anthracite to transition rocks, and the graphite to primitive rocks. I do not know any limestone formation in the Andes which resembles those contained in the group § 20. I saw at Contreras only, at the eastern foot of the Cordillera of Quindiu (New Grenada), a transition limestone, not compact, but very granular, bluish grey, mixed with grains of quartz, and containing siliceous masses resembling pitchstone. These masses are traversed by veins of calcedony. The position of this limestone of Contreras, in the midst of a formation of sandstone and secondary gypsum, is difficult to determine.

II. TRANSITION PORPHYRY AND SYENITE IMMEDIATELY COVERING PRIMITIVE ROCKS, BLACK LIMESTONE, AND GREENSTONE.

§ 21. This is the great formation, destitute of grauwacke, of South America. It presents some problems very difficult to solve, and comprehends the transition porphyries of the Andes of Popayan, and of that part of Peru which I passed over in returning from the Amazon river to the South Sea. Previously to any detailed description of this formation, I shall take a general view of the porphyroid rocks of equinoxial America, which have been the principal object of my geognostic researches. If, as M. Mohs has well observed, grauwacke characterises, pre-eminently, in Germany and a great part of Europe, the intermediary formations, we may in the equinoxial region of the new continent regard porphyries as the principal type of those formations. No other chain of mountains contains a greater mass of porphyries than the Cordilleras, which extend, almost in the direction of a meridian, 2500 leagues from one hemisphere to the other. These porphyries, in part rich in ores of gold and silver (§ 23.), are most frequently associated with trachytes, by which they are covered, and through which the volcanic agents still penetrate. This association of metalliferous rocks with rocks produced or changed by fire would less astonish the geognosts of Europe, if it did not extend to gold and silver, but only to specular (fer oligiste) iron, oxidulated iron (fer oxidulé) titaneous iron, and muriate of copper. This phenomenon is one of the most striking and most opposed to the opinions long entertained by celebrated men. It is, however, a fact very necessary to be well determined, that there is a proximity of position, and sometimes an analogy in the composition, without an identity of formation. The method we have adopted of circumscribing different formations according to their superposition, and the nature of the rocks by which they are covered, will, I flatter myself, serve to throw some light on the relations which are observed between transition porphyries, trachytes, and (secondary) porphyries of the red sandstone. I shall, at the same time, indicate the places where limits, as strongly marked as the actual state of our systematic divisions seem to require, have not yet been discovered in nature.

The porphyries of South America may be considered in two ways; according to their geographical position, and according to the dates of their formation. We find in Europe, transition porphyries and syenites (Saxony, Vosges, Norway,) generally distant from trachytes (Siebengebirge, near Bonn, Auvergne); it happens, however, also, that porphyries and trachytes are found united (Hungary); and then the former are sometimes metalliferous. In South America, porphyries and trachytes are all found together on a narrow land, in the most western and most elevated part of the continent, on the shore of that immense basin of the Pacific ocean which is bounded, on the side of Asia, by the volcanoes and trachytic rocks of the Kuriles, Japanese, Philippine, and Molucca Islands. At the east of the Andes, throughout the whole eastern part of South America, on a space of ground of more than 500,000 square leagues, no transition porphyry, nor real basalt with olivine, nor trachyte, nor burning volcano, have been observed either in the plains or the groups of insulated mountains. The phenomena of the trachyte formation appear to be confined to the ridge and the line of the Andes of Chili, Peru, New Grenada, Saint Martha, and Merida. I announce this circumstance in a particular manner, in order that travellers may be induced to confirm it by farther examination, or refute it. In the same region, which extends from the eastern declivity of the Andes towards the coast of Guiana and Brazil, gold, platina, palladium, tin, and an immense quantity of specular and magnetic iron have been found; but amidst many indications of sulphuret or muriate of silver, no mine has been discovered which can be compared in richness to those of Peru and Mexico. I did not see transition porphyries, nor the porphyries of red sandstone, in the chain on the coast of Venezuela, in the Sierra de la Parima, nor in the plains between the Oroonoko, the Rio Negro, and the Amazon river. To the east of the Andes, I know but one small portion of the trachyte formation near Parapara (northern border of the Llanos of the Caracas), where, in a spot highly interesting to geognosy, phonolite and mandelstein with pyroxene are superposed on transition serpentines and clay-slates; but these phonolites are found on the border of the chain of the Cordillera of Caracas, which is connected by Nirgua, Tocuyo, and the Paramo of Niquitao, to the Andes of Merida. M. d'Eschwege found, in Brazil, some porphyries interposed by beds in primitive formations of granite-gneiss; but he thinks, also, that this vast country is destitute of independent formations of transition porphyry, trachyte, basalt, or dolerite. The very great length of the rivers in America, and the number of their tributary streams, facilitate, by the examination of rolled stones, the knowledge of those parts of the country which have not been visited. I collected between Carare and Honda, in the midst of a formation of sandstone, fragments of trachytes which the river Magdeleine receives from the Andes of Antioquia and Herveo (New Grenada).

With respect to the nature of the formations of porphyry which exist so abundantly in the western and mountainous land of South America, and in that of Mexico, which is but a prolongation of the same land, we shall describe two very distinct groups in that place. The first (§ 21.), not metalliferous, reposes immediately on primitive rocks; the second (§ 23.), often metalliferous, rests on clay-slate, or on talcose slate, with transition limestone; both of these by their position and composition sometimes resemble trachytic porphyries, as the porphyries of the group § 22. resemble those of the red sandstone. In fact, the transition porphyries of the Andes of Peru and Mexico are often found covered by trachytes, while the porphyries of some parts of Germany are covered by the secondary formation of red sandstone, which contains in its turn porphyries and mandelstein. In equinoxial America, the limits between transition porphyries and real trachytes known to be volcanic rocks are not easy to fix. In ascending from the porphyries which contain the rich silver mines of Pachuca, de Real del Monte, and of Morau, (porphyries destitute of quartz, but often abounding in hornblende and common feldspar,) towards the white trachytes with pearl-stone and obsidian of Oyamel, and of Gerro de las Navajas (mountain des Couteaux, to the east of Mexico); and in passing, in the Andes of Popayan, transition porphyries covered on some points with fine-grained black limestone, to the pumice-trachytes that surround the volcano of Puracé, we find intermediary porphyritic rocks, which we are tempted sometimes to regard as transition porphyries, sometimes as trachytes. To this may be added, that amidst these porphyries of Mexico so rich in gold and silver, we observe beds (Villalpando near Guanaxuato,) destitute of hornblende, but containing slender crystals of glassy feldspar. They cannot be distinguished from the phonolites (porphyr-schiefer) of Biliner-Stein, in Bohemia. Generally, as the learned professor of mineralogy at Mexico, M. Andres del Rio, one of the most distinguished pupils of the school of Werner, had observed before me, the transition porphyries of New Spain contain two species of feldspar, the common and the vitreous. It appeared to me, that the latter becomes more abundant in the upper beds, in proportion as we approach the trachyte-porphyries.

In the equinoxial part of the new continent, we are as much embarrassed by the connection of the porphyries, often argentiferous, with the trachytes containing obsidian, as we are in Europe by the close connection of the last transition rocks with the most ancient secondary rocks, or the alternation of transition mica-slates, which have every appearance of primitive rocks with grauwacke and very ancient conglomerates. The source of this embarrassment is not, however, the same. There is nothing extraordinary in seeing fragmentary rocks, or rocks containing orthoceratites, madrepores, and encrinites, succeeded by rocks destitute of organic remains, and resembling gneiss and primitive mica-slate. This alternation, this local and periodical absence of living beings, manifests itself even in secondary and tertiary formations; it appears to indicate different states of the surface of the globe, or of the bottoms of the basins in which the stony deposits have been formed. On the contrary, the association of transition porphyries and trachytes, and the frequent passage of these rocks into each other, are phenomena that seem to loosen the bases of those geogonic principles which have been most generally received. Must we consider trachytes, pearlstone, and obsidian, as of the same origin with clay-slates containing trilobites, and black limestones with orthoceratites? Or ought we not rather to admit, that the domain of volcanic action has been too much limited; and that these porphyries, partly metalliferous, destitute of quartz, but containing hornblende, vitreous feldspar, and even pyroxene, are with respect to their origin and relative age connected with trachytes, as these trachytes formerly confounded with transition porphyries by the name of trap-porphyries are connected with basalts and real lava ejected by burning volcanoes? The first of these hypotheses appears to me to disagree with all that has been observed in Europe, and all that I have been able to collect respecting the obsidian and pearlstone at the peak of Teneriffe, and at the volcanoes of Popayan and Quito. The second hypothesis will appear less hazardous, less devoid of probability, perhaps, when we no longer limit the idea of volcanic action to the effects produced by the craters of our burning volcanoes; and when we consider that action as owing to the high temperature which every where prevails at great depths in the interior of our planet. We have seen within the period of history, even in that part nearest to our time, rocks of trachyte rise out of the sea without flames, without any ejection of scoria (Archipelago of Greece, the Acores, and Aleutiennes isles); we have seen balls of basalt, in concentric layers, issue from the earth completely formed, and collect together in small cones (Playas of Jorullo, in Mexico). Do not these phenomena lead us to imagine to a certain degree what may have formerly taken place on a much greater scale in the fissured crust of the globe, wherever that internal heat which is independent of the inclination of the axis of the earth, and the slight influences of climate, has, through the intervention of elastic fluids, heaved up masses of rocks more or less softened and liquefied?

In mentioning the transition formations, which, in the Andes of Mexico, New Grenada, and Peru, seem to be connected with the trachytes by which they are covered, we cannot avoid entering into some considerations on the origin of rocks. It is the imperfection of our classification of formations which leads to this digression. The word *vol-*

canic rock denotes, as I have observed above, a principle of division quite different from that which we follow in separating primitive from secondary rocks. In the latter case, we indicate a fact susceptible of direct observation. Without advancing further, and in examining only the actual state of things, we can decide if an association of rocks is, or is not, entirely destitute of organic remains, or whether or not any arenaceous or fragmentary beds are found interposed. On the contrary, in opposing volcanic formations to primitive and secondary formations, we agitate a question *altogether historical;* we oblige the geognost to pronounce against his will as by exclusion, on the origin of granites, syenites, and porphyries. It is no longer the direct observation of what exists, the presence or the absence of impressions of organised bodies; it is by reasoning founded on inductions and analogies more or less contested, that he is obliged to decide on the *volcanic* or *non-volcanic* origin of a formation. Among the products which the greatest number of geognosts, I may say all who have seen Italy, Auvergne, the Canaries, and the Andes, consider as decidedly igneous (porphyries with a base of obsidian, semi-vitreous porphyries, trachyte-porphyries, and the porphyries which by their composition, by the presence of quartz, by the absence of vitreous feldspar, hornblende, and pyroxene, resemble the grauwacke-porphyries), beds are found (in the Cordillera of the Andes) of which the base passes to phonolite (the base of porphyr-schiefer), and in which vitreous feldspar, hornblende, and sometimes even pyroxene, progressively replace common feldspar. We do not then know where the porphyries which we are agreed to call transition end, and where the trachytes begin.

I have no doubt that new travels, and a profound examination of intermediary feldspathic rocks, and of those contained in the red sandstone, will throw more light on this interesting problem; in the present state of our knowledge, I shall content myself to be guided in the separation of the porphyries and trachytes of the Andes, less by the consideration of their composition than by that of their position. It is extremely unusual to find common feldspar in the real trachytes of equinoxial America; but vitreous feldspar, hornblende, and pyroxene, are found in these rocks, and in the porphyries, (§§ 21. and 23.), which are partly covered by a black transition limestone and by secondary red sandstone. We find also but little quartz in the porphyries and trachytes of equinoxial America; on the contrary, that substance characterises the greater part of the porphyries of Europe (§§ 22. and 24.). Its total absence, however, is not a certain indication of a trachytic formation, as it occurs, although in small masses, in some trachytes of the Dardanelles, Hungary, and Chimborazo. M. de Buch has observed, near the basalts of Antrim, a porphyry very analogous to those of the red sandstone; and containing quartz, common feldspar disseminated, together with interposed beds of pearlstone and obsidian. This phenomenon also occurs in the trachytes of the Euganean mountains. Mica, and also garnets, appear, although very rarely, in the transition porphyries of both continents; but they occur equally in the trachytes of the ancient volcano of Yanaurcu, at the foot of Chimborazo, and in the trachytic conglomerate of Europe. The porphyries, as well as the trachytes of the Andes, exhibit superb columns; and the mass of columnar trachyte is sometimes so compact, that it is difficult to discover in them either pores or fissures.

From these statements it results, that the characters of composition (absolute and insulated characters, by which some have attempted to distinguish the transition porphyries, and the trachytes of the Cordilleras) are very uncertain; it is by the whole of these oryctognostic characters, by the passage of a rock to a vitreous state, by the obsidian, the pearlstone and the scorified masses which are imbedded in it, and by the relations of position, that a trachyte can be known. Besides, it is easier to class certain formations as trachytes, than to decide on the pretended neptunian origin of other rocks. Trachytes and transition porphyries may both be placed on primitive rocks; it is not the rocks by which they are supported, but those by which they are covered, that ought to guide the geognost. The trachytes and porphyries of the Cordilleras are, most frequently, not covered by other formations; but wherever they are so covered, and where the superposed rock is indubitably transition, that superposition alone determines, in my opinion, the problem of classification. Trachytes generally serve as a basis only to other igneous products; very rarely (Hungary) to tertiary formations identical with those round Paris; still more rarely (Archipelago of the Canaries, and the Andes of Quito) to thin formations of gypsum and oolites interstratified with, or superposed on, volcanic tufas. The transition porphyries of America, and not the trachytes, are sometimes covered by fine-grained black limestone, by red sandstone, or alpine limestone; and it is when this covering is wanting that we are obliged to have recourse to the uncertain method of inductions and analogies. We should, perhaps, risk less in separating what nature has united closely, if provisionally we described, under the vague denomination of *amphibolic* porphyries (hornblende-porphyries, hornblendiges porphyrgebilde), the whole of those rocks of the Cordilleras which exhibit a porphyroid structure (transition porphyries and trap-porphyries, or trachytes), which are almost destitute of quartz, and which abound at the same time in hornblende and in lamellar or vitreous feldspar.

After having given this general view of the transition porphyries of the Andes, and their geognostic affinity with trachytes, I shall describe that group of porphyries which is anterior to limestone with entrochites and orthoceratites, and to clay-slate and transition mica-slate. In this equatorial group, we may distinguish, in the northern hemisphere (Cordilleras of Popayan and Almaguer), and in the southern hemisphere (mountains of Ayavaca, on the limits of the Andes of Quito and Peru), several partial formations; viz: —

Porphyries;
Greenstone and ferruginous clays;
Syenites;
(Transition granites?);
Limestone containing much carbon;
(Transition gypsums?).

Porphyries, the aspect of which is often trachytic, predominate in this group. I did not observe porphyries alternating with syenite, or with transition limestone; nor syenite with greenstone, which occurs (§§ 23. and 24.) at Mexico, and in several parts of Europe. The syenite of the Andes of Baraguan, Chinche, and Huile (to the east of Rio Cauca, between Quindiu and Guanacas, north lat. 2° 45′ to 4° 10′), is placed on primitive rocks, on granite-gneiss, and perhaps even on mica-slate. It is a partial formation which is parallel to the porphyries of Popayan, and covered by limestone with much carbon. This syenite is composed of a large quantity of hornblende, and common reddish-white feldspar, with very little black mica and quartz. Feldspar predominates in the mass; the quartz (which is remarkable in a syenite) is translucid, whitish-grey, and constantly crystallised, like the quartz of the porphyries of Europe.

§ 24. The aggregation of the parts is almost in layers, so that the transition syenite of the Cordilleras is not of a texture entirely granular, like the syenite of Plauen, near Dresden; the texture (*flasrige structur*) of this rock approaches, on the contrary, to that of gneiss. What removes the syenite of Nerado de Baraguan from granites with hornblende (§ 7.), or from a syenite which we might consider as primitive (§ 8.), is its passage to trachyte, and its connection with the transition greenstone which is superposed on it, between the Paramo d'Iraca and the Rio Paez (province of Popayan). Quartz disappears gradually in this transition syenite, hornblende becomes more abundant, and the rock assumes the porphyroid structure. We then find in a petrosiliceous paste (euritic) of a reddish or yellowish-grey colour, very little black mica, a great quantity of hornblende and disseminated slender crystals of feldspar with rather a vitreous than pearly lustre, the laminæ of which, though not distinct, have longitudinal fissures. It is no longer a syenite, but a trachyte, of which the numerous masses, variously grouped, rise like fortified castles, on the ridge of the Andes. These passages appear to me very remarkable, and seem to confirm the doubts which may be entertained on the origin of all primitive granular rocks. It is very difficult, in equatorial countries, to apply names to a great number of mixed formations of feldspar and hornblende, because those formations are found just between the transition syenites and the trachytes. Sometimes granular, sometimes porphyroid, they resemble either the syenites of the group (§ 23.) of Hungary, or the trachytes of Drachenfels, near Bonn, and of the great table-land of Quito. It being observed that the transition porphyries of Popayan pass also to trachytes, the parallelism of formation between the syenites and porphyries of the same group (§ 21.) is confirmed by the geognostic relations of two rocks with a third. Sometimes (foot of the volcano of Puracé, near Santa Barbara,) a *transition granite*, very abundant in mica, seems to separate the syenites (in which are imbedded quartz and common feldspar with a pearly lustre) from real trachytes, of which the paste, towards the summit of the mountains (at the height of 2,200 toises) becomes vitreous, and passes to obsidian.

In the whole group of syenites and porphyries which I have examined in the Cordilleras of the Andes (between Nevado de Tolima and the towns of Popayan, Almaguer, and Pasto), the porphyry which bears most decidedly the character of a transition rock is that which surrounds the basalts of Tetilla de Julumito (left bank of the Rio Cauca, west of Popayan), and which is covered (at Los Serillos) by a *blackish limestone*, passing from compact to small-grained, traversed by veins of white calcareous spar, and containing so much carbon that in some parts it blackens the fingers; and is even found in powder in the clefts of stratification. This accumulation of carbon, which is also observed in anthracitous and aluminous slates, and in lydian stone, and kiesel-schiefer, leaves no doubt respecting the question, whether the darkish limestone of Los Serillos (near Julumito), in which I could find no trace of organised remains, is a real transition limestone. The lydian stone which is observed in the transition clay-slate of Naila and Steben (mountains of Bareuth) furnishes also this deposit of carbonaceous powder between the clefts; and I employed specimens which do not stain the finger to excite the nerves of a frog, by placing them in a galvanic circle, conjointly with zinc. The black transition limestone *(nero antico)*, so celebrated among the ancients by the name of *marmor Luculleum*, contains also, according to the analysis of M. John, ¾ per cent. of oxide of carbon, distributed as a colouring principle through the whole mass of the rock. Porphyry, covered by a highly carburetted limestone, greyish-black, fine grained, and perhaps destitute of petrifactions, will be considered by the geognost who attaches more importance to the position than to the composition of formations, as a transition porphyry, whatever may be oryctognostic nature of its constituent parts. Trachytes, as we have observed above, have hitherto been found covered only by other volcanic rocks, tufas, or some very recent tertiary formations. The transition porphyry of Popayan, on which black limestone is placed, is rather regularly stratified; it contains little hornblende, very little quartz in small crystals imbedded in the mass, and feldspar which passes from common to vitreous. I saw no pyroxene there, nor in the porphyry of Pisojé which forms a magnificent colonnade at the western declivity of the volcano of Puracé, on the right bank of the Rio Cauca. This porphyry of Pisojé is divided into prisms, of from five to seven sides, and eighteen feet long, which I mistook at a distance for basalt, and which are found in Europe in many transition porphyries, even in those of the red sandstone. A perpendicular range of these columns is placed on a range entirely horizontal. In a greenish-grey paste, probably compact feldspar coloured by hornblende, there are found a very few crystals of hornblende visible to the naked eye, black mica, and a great quantity of milky and non-vitreous feldspar. Quartz is wanting in these columnar porphyries, as in almost all the transition and metalliferous porphyries of Mexico. The rock of Pisojé being considerably distant, geographically, from the porphyries of Julumito connected with transition limestone, it remains doubtful whether it does not belong to a trachyte formation. With respect to the transition porphyries of Julumito, we do not know on what formation they repose; since, from Quilichao as far as the ridge of Los Robles, which is situated at the west of Paramo de Palitarà and the volcano of Paracé, and which divides the waters of the South Sea and the Sea of the Antilles, no primitive rocks appear. The Alto of Los Robles itself is composed of mica-slate (direction of the beds N. 60 E. like the micaceous gneiss of the Andes of Quindiu; inclin. 50° to S.W.). This primitive rock of the Robles is also observed near Timbio, and near the sources of the Rio de las Piedras (height 1004 toises) which issue from below the trachytes of Paracé and Sotara. I saw clearly that in the ravines between the Rio Quilquasé and Rio Smita, the porphyritic rocks of Cerro Broneaso, and those which follow towards the south between Los Robles and the Paramillo of Almaguer, repose on mica-slate. The great blocks of quartz, therefore, that are found scattered amidst these porphyry and trachyte formations, denote every where the proximity of mica-slate.

Here an important question arises, whether the rocks of porphyroid structure, at the south of the Alto de los Robles, forming the western declivity of the volcano of Sotara and the Paramos de las Papas and Cujurcu (see my map of the Rio Grande de la Magdalena), are true transition porphyries? I shall state the facts which I observed. The porphyries of Broncaso (north lat. 2° 17′, long. 79° 3′, deducting this position from the astronomical observations which I made at Popayan and at Almaguer) contain many and very large crystals of milk-white feldspar, slender crystals of hornblende that cross each other as the feldspar in the porphyry commonly called by the antiquarians *serpentino verde antico* or *porfido verde* (grünporphyr of Werner), and a little translucid crystallised quartz. The crystals of hornblende and feldspar often divide from the same point. In the interior of the feldspar we find other crystals, very small and black, which I thought were rather pyroxene than hornblende. The central point around which the crystallised laminæ of leucite (amphigène) are grouped, is equally, according to M. de Buch, a microscopic crystal of pyroxene; and in the pophyritic greenstones of Hungary, M. Beudant found garnets among crystals of hornblende. Singular crossings and groupings in the crystals of common feldspar and of hornblende characterise all the porphyries between the Cerro Broncaso, and the valleys of Quilquase and Rio Smita, porphyries which are irregularly stratified in non-conformable stratification (beds of from two to three feet; direction N. 55° W.; inclin. 40° to N.E.) with beds of mica-slate. Their paste differs from that of the porphyries of Julumito; it is of a fine asparagus green, of a compact or scaly fracture, sometimes very fragile, presenting a grey streak, and assuming, when breathed upon, a very deep colour; at other times it is hard, and resembles jade or phonolite (klingstein, base of porphyr-schiefer), that is, it belongs to compact feldspar. I saw on the banks of Rio Smita, in those porphyries which pass to the *porfido verde* of the antiquaries, beds having very few disseminated crystals; they are masses of jade (saussurite), asparagus and leek-green, nearly similar to that which occurs in the transition euphotide; they are traversed by an infinity of small veins of quartz. Further south, the green porphyries with a base of compact feldspar retain their disseminated crystals of quartz; and this character separates them from the porphyry-slate belonging to the trachytic formation, in which quartz is a phenomenon of rare occurrence. At the same time black mica begins to appear, and a variety of pyroxene having a very brilliant surface, the transverse fracture conchoidal, of an olive-green colour, and so light that it might almost be taken for the olivine of basalt. This porphyry with black mica fills the valleys of the small rivers of San Pedro, Gauchicon, and Putes; it is sometimes concealed (valley of La Sequia) beneath masses of greenstone, in balls from four to six inches in diameter; and finishes by being no longer stratified, but separated, exactly like the superposed greenstone, into balls which divide by decomposition into concentric layers. The balls of porphyry, often extremely hard, are of a composition identical with the porphyry in mass. Their nucleus is solid, and contains neither quartz nor calcedony; they form separate strata of six feet in thickness, and are found as if imbedded and fixed in the rock, not altered by atmospheric or galvanic influences. This structure is not an effect of the decomposition, similar to what has been thought, of some columnar basalts that separate into balls. It appears to me rather to be connected with a primitive arrangement of moleculæ. I believe there is no where to be found a greater quantity of rocks having a *globular structure* than in the Cordilleras of the Andes, and principally from Quilichao (between Caloto and Popayan) as far as the small town of Almaguer.

In descending the Cerro Broncaso, and crossing successively (in the direction from north to south, and in the way from Popayan to Almaguer) the valleys of Smita, San Pedro, and Guachicon, we observe in the midst of a porphyry not divided into balls, and containing more hornblende and more olive-green pyroxene than vitreous feldspar, a very remarkable geognostic phenomenon. Angular fragments of gneiss, from three to four inches square, are imbedded in the mass. It is a gneiss abounding in mica; a phenomenon which the trachytes of Drachenfels present, (also Siebengebirge, on the banks of the Rhine,) and, in its lower beds, the phonolite (porphyr-schiefer) of Biliner Stein, in Bohemia. Not far from thence, in the north-east part of the same valley of Rio Guachicon (a valley 400 toises in depth, where I stopped during a whole day), the porphyroid rock has the most complicated structure which I have ever seen in transition porphyries, or porphyritic trachytes. We may there observe at the same time vitreous crystals of feldspar, hornblende, black mica, quartz, and pyroxene the colour of which approaches that of olivine. The quartz occurs only in very small masses, but these are certainly not owing to posterior filtrations. After having passed still further to the south the ridge that separates the Rio Guachicon from the Rio Putés, the five substances disseminated in the mass disappear almost entirely; the porphyroid rock becomes homogeneous, extremely hard, and of that fine black which is admired in some very pure lydian stones, or in the basis of the pretended porphyritic jasper of Altaï, or in some Egyptian statues falsely called *basaltes* or *basanites*. I doubt its being pechstein; it is rather a compact feldspar coloured black by hornblende or some other substance. The fracture of this homogeneous paste is smooth or conchoidal, with large flattened cavities; it is almost entirely without lustre. I observed only a few very long crystals of vitreous feldspar, and hexahedral prisms of conchoidal pyroxene (muschligen augit of Werner), which have the black colour of melanite, and resemble in their lustre and fracture the pyroxene of Heulenberg, near Schandau in Saxony.

I have now described successively the porphyries of Julumito, which are covered by black carbonate of lime; those of Pisojé, with common feldspar, and divided into prisms; green porphyry, containing quartz, and frequently macled crystals of hornblende, from Cerro Broncaso, and the valley of Smita; the porphyroid rocks of Rio Guachicon, enclosing fragments of gneiss; finally, those of Rio Putés, of which the homogeneous and compact black mass contains but few disseminated crystals. Do all those rocks belong to the same formation, which presents peculiar characters in the different valleys of the Cordillera of Sotara and of Cujurcu? It cannot be doubted that the fragments of gneiss imbedded in the rocks around the Rio Guachicon characterise real trachytes. They are in a manner precursors of those trachytes, and that enormous mass of pumice which I found, twenty leagues further south, on the banks of the Mayo. But must this denomination of trachyte be extended to all the porphyries that stretch by the Cerro Broncaso, towards the mica-slates of the Alto de los Robles, and which are partly covered, not by dolerite, but by greenstone of a globular structure resembling the transition greenstone of Germany? After what I stated above on the insensible passage of the metalliferous porphyries of Mexico to rocks that contain obsidian and pearlstone, of which the volcanic origin is now scarcely contested, I know not how to decide on a question of so much importance. It does not present so much a problem of position as a problem which may be called *historical*, because it is the object of geogony, and is intimately connected with the ideas which we form on the origin of those various rocky deposits that cover the surface of the globe. When the geognost has examined the relations of position and composition, he has fulfilled his task. It is not yet the time to pronounce respecting those masses that seem to oscillate between transition porphyries and trachytes, called exclusively volcanic porphyries. What is now difficult to unravel will perhaps become clear, when equinoxial America, free, civilised, and more accessible to travellers, shall be explored by many well-informed men; when, from new discoveries, it may be conceived that volcanic effects, whether slow and progressive, or rapid and tumultuous, may have taken place wherever fissures have opened communications with the interior of the globe, in which, according to every appearance, a very elevated temperature still prevails. We have already certain proofs that rocks almost identical with those that belong to the trachytic formation, or by which they are covered, are interposed in real transition porphyries, and in the porphyries of the red sandstone. Geognosts are well acquainted with the important observations made by M. de Buch, near Holmstrandt, in the gulf of Christiania in Norway. A porphyry containing, besides common feldspar (not vitreous), a little hornblende and quartz, is found placed between a limestone with orthoceratites and a syenite with zircons. No one has yet hesitated to consider this porphyry as a transition formation; no one has called it trachyte. But in the midst of this porphyry we see, not a vein (dyke), but a bed of basalt with pyroxene. "The porphyry of Holmstrandt," says M. de Buch, "becomes basalt by the same passages and insensible gradations which we find so commonly in Auvergne. This basalt is very black, almost fine-grained, destitute of feldspar, but filled with pyroxene. Sometimes it becomes cellular, and assumes a red and scorified aspect, at the contact with porphyry." It would not be more extraordinary, perhaps, to discover fragments of gneiss enveloped in this cellular and scorified basalt filled with pyroxene, than to find them in the basalts of Bärenstein (near Annaberg in Saxony), or in the trachytes of the valley of Rio Guachicon (in South America). What is the origin of this balsatic, cellular and pyroxenic bed of Holmstrandt? Has it been, as well as all porphyry, a stream that issued from below as veins? Does the presence of a mass which is supposed to be of igneous origin, afford a sufficient reason for admitting that the whole formation to which this mass belongs should be separated from the transition formations, and classed among trachytes? This I doubt; the incontestably volcanic rocks of Rio Guachicon, containing fragments of gneiss, are geognostically connected with transition porphyries, as these are, on other points of the globe, geognostically allied to the porphyries of the red sandstone.

I separate provisionally all the porphyroid rocks placed to the south of a ridge composed of mica-slate (Alto de los Robles) from those which are found at the north-west of that ridge, and which, near Julumito, are covered with limestone abounding in carbon. To the latter class, and consequently to the transition formation (§ 21.) which forms the particular object of this article, I assimilate, perhaps with more confidence, the porphyries of Voisaco (Andes de Pasto, north lat. 1° 24′) and those of Ayavaca (Andes of Peru, south lat. 4° 38′). The following are the circumstances respecting the position of those two rocks. The porphyries and trachytes of Popayan, of Cerro Broncaso, Rio Guachicon, and Rio Putés, are separated from those of the province of Pasto by a table-land of primitive rocks, extending from Almaguer as far as Tablon, at the foot of the Paramo of Paraguay. The porphyries recommence at the south of Tablon; near the Indian village of Voisaco they are distinguished by a polarity which is found sensible even in the smallest fragments. These porphyries, it may be clearly seen, are placed on mica-slate. A greenish-grey mass encloses at the same time two varieties of feldspar, the common, and the vitreous; a phenomenon which is often found in the transition porphyries of Mexico (§ 25.). Some acicular crystals of pyroxene penetrate between the laminæ of vitreous feldspar. A rock placed at the entrance of the village presented to M. Bonpland and myself all the phenomenon of the magnetic serpentine of Bareuth (§ 19.), which I had discovered in 1796.

In the northern hemisphere, following the Andes of Quito by Loxa to Ayavaca, we see the primitive rocks and porphyries appearing alternately, a phenomenon which we have already mentioned (§§ 5. and 6.). Almost every time that the mass of the mountain rises, the porphyries appear, and conceal from the eye of the traveller the gneiss and mica-slate. Those porphyries, which at first present more of common than of vitreous feldspar, are succeeded by trachytes which usually announce two combined phenomena, the vicinity of some volcano still burning, and the rapidly increasing elevation of the Cordillera, the summits of which nearly attain or overpass the limit of perpetual snow (2460 toises under the equator). I shall add that trachytes cover immediately either primitive rocks or transition porphyries, and that vitreous feldspar, hornblende, and sometimes pyroxene, become more frequent in the latter, in proportion as they are found nearer to volcanic rocks. Such is the type which the phenomena of position present in the equinoxial region of Mexico and South America; a type which I have particularly adhered to in the geognostic profiles which I drew on the spot in 1802 and 1803.

The porphyries of Ayavaca form a part of this general chain of feldspathic rocks. On the mica-schists of Loxa, where are the finest trees of Jesuit's bark hitherto known (*Cinchona condaminia*), porphyries are placed that fill the whole space comprehended between the valleys of Catamayo and Cutaco, near Lucarque and Ayavaca (1407 toises high). These porphyries are found divided into balls with concentric layers, and heaps of those balls lie (valley of Rio Cataco, height of the bottom of this ravine, 756 toises) on a porphyry which contains common feldspar and hornblende which is regularly stratified, and the mass of which, of great density, is traversed by an infinity of small veins of carbonate of lime, as transition clay-slate in Europe is traversed by veins of quartz. The barometrical measures which I have taken, assign 4800 feet of thickness to those porphyries of Ayavaca, which I do not consider as trachytes. I do not mention as belonging to the group § 21., the green porphyroid rocks destitute of quartz, containing very little hornblende, and a great deal of common milk-white feldspar, which constitute the Andes of Assuay. They are placed on the primitive mica-slates of Pomallacta, and I had the opportunity of examining their enormous thickness, from 1500 to 2074 toises of height above the level of the ocean. They are generally stratified; but this stratification, often very regular (N. 45° W.), is observed also in many real trachytes of Chimborazo and the burning volcano of Tunguragua. In carefully examining, in the Cordilleras of the Andes, the different states of feldspar in transition porphyries and trachytes, I observed that rocks decidedly trachytic, also contain some that are not vitreous, but laminar and milky. I am inclined to think that the porphyry of Assuay, a group of mountains celebrated for the passage it affords between Quito and Cuença, is of trachyte.

I have discussed the nature of the rocks which in South America constitute the group § 21., the syenite of Baraguan, the transition granite of Santa-Barbara, the porphyries of Julumito, the greenstone, and black and carburetted limestone: it remains for me to make some observations on the less important members of this group. The salt springs which are found surrounded by syenite at an immense height near San Miguel, at the east of Tulua, in the Cordillera of Baraguan, perhaps indicate the geognostic connection of some transition gypsum with syenite, or with a black limestone analogous to that of the Serillos of Popayan. But in those countries height alone does not afford a reason for excluding a gypsum formation from the domain of secondary formations. I saw on the table-land of Santa Fé de Bogota, at 1400 toises high, the rock salt of Zipaquira reposing on a limestone which is decidedly of secondary formation. It is rather probable that the fibrous gypsum mixed with clay of Ticsan (Pueblo viejo, in the kingdom of Quito, south lat. 2° 13′.), situated opposite to the famous mountain of sulphur (§ § 11. and 16.), far from every secondary rock, or primitive mica-slate, is a transition gypsum analogous to that of Bedillac in the Pyrenees, and to that of Saint-Michel near Modane in Savoy.

The greenstones of group § 21. which appear to cover the syenites of Baraguan, and porphyries analogous to those of Julumito abound at the north of Popayan, at the foot of the Paramos of Iraca and Chinche, and, chiefly, in the eastern valley of the basin of Rio Cauca (Curato de Quina Major and Quilichao). Rich gold-washings are carried on in the latter spot, among the fragments of greenstone (diabase of Brongniart, diorite of Haüy). This rock is certainly not a dolerite, but a transition greenstone, similar to that which is found interposed in the clay-slate impregnated with carbon of the Fichtelgebirge (§ 22.), and in the mica-slate of the Caracas (§ 11.). The greenstone of Quina Major becomes sometimes extremely black, very homogeneous, sonorous, fissile, and stratified, like the hornblende schist of primitive formations (hornblend-schiefer). It is filled with pyrites, does not act upon the magnet, and acquires a yellowish coat in the air, like basalt. Near Quilichao (between the towns of Cali and Popayan) it presents large crystals of hornblende disseminated in the mass, and veins which are filled with pyroxenes of an olive-green colour, not dark. On the spot, I took these pyroxenes for the lamellar olivine of M. Freiesleben. The crystals are not disseminated in the mass, but occur in the fissures, like the veins of dolerite that traverse the greenstone. This rock, although destitute of veins, appears, as we have said above, in flattened balls, at the south of Popayan and the Alto de los Robles. In the valley of La Sequia (between the Cerro Broncaso and the Rio Guachicon) it covers the green porphyries of Rio Smita. The superposition of the greenstone is here more manifest than in the Curato of Quina Major, and in the gold-washings of Quilichao. The porphyries at the north of the Alto de los Robles being partly (Julumito) covered with black transition limestone, and those which we observe at the south of Los Robles being connected with the trachytes of Rio Guachicon, this uniform superposition of greenstone on both these porphyries is a phenomenon of position that merits great attention. According to the observations hitherto made in both continents, trachytes and basalts are found covered by dolerite (an intimate mixture of feldspar and pyroxene), but not by greenstone (an intimate mixture of feldspar and hornblende).

Must we not thence conclude that all which is below the greenstone in balls of Sequia and Quilichao is a transition porphyry, and not a trachyte? Ought we not, on account of this uniform superposition of greenstone, to separate the porphyroid rocks of Rio Smita and the Cerro Broncaso from the trachytic porphyries of the valley of Guachicon, which are more decidedly pyroxenic, that is, from those in which fragments of gneiss are imbedded? There is some probability that a rock covered by greenstone is rather a transition formation than one of trachyte; but formations of igneous origin may be of a very ancient date. Why should not masses of trachyte and dolerite be interposed in the newer transition rocks?

Further—and I put this question to the learned mineralogists who have devoted themselves more especially to the study of the oryctognostic characters of rocks,—are greenstones always as different from dolerites mineralogically (by their composition) as they are geognostically, by their position? The substance of crystals that are insulated in a paste, and which become visible to the naked eye, no doubt exists mixed with other substances in the paste itself. As basalts often contain at the same time (Saxony, Bohemia, Rhunegebirge,) large disseminated crystals of pyroxene and hornblende (basaltische hornblende), it cannot be doubted, that besides the pyroxene, hornblende also enters into the mass of some basalts. Why might not analogous mixtures take place in the paste of dolerites and greenstones, of which the one (to use the mythologic nomenclature generally received) is believed to be of volcanic, and the other of neptunian origin ? The pyroxene rock, which, according to M. Charpentier, is found in parallel stratification in the primitive limestone of the Pyrenees, contains disseminated hornblende. It is asserted, that pyroxenes have been observed in greenstones which form true beds amidst the granites of the Fichtelgebirge in Franconia (§ 7.). M. Beudant saw greenstone indubitably pyroxenic (consequently dolerite) in the transition porphyries and syenites of Hungary (Tepla, near Schemnitz), as well as in the coal-sandstone (secondary) of Funfkirchen. The stratified and globular greenstone in the vicinity of Popayan passes neither to mandelstein nor to syenitic porphyry. It is a very marked formation, and accompanied here, as it is almost every where in the Cordillera of the Andes (where it occurs at some distance from the chain of active volcanos), by enormous masses of clay. These masses remind us more of the accumulations of clay in the basaltic formations of Mittelgebirge in Bohemia, than of the clays belonging to the gypsum of the greenstone (uphites of Palassou) in the Pyrenees, and the department of the Landes. They render the passage of the Cordilleras, from Popayan to Quito, extremely difficult during the rainy season.

The analogies which we have indicated between some porphyries of the group § 21., and the trachytes, or other volcanic rocks, is found in the

Mexican group, § 23., and even in Norwegian porphyries of the group § 24.; but generally (with the exception of the porphyries of Caucasus) they are very seldom observed in the porphyries subordinate to transition clay-slate and grauwacke, § 22. We may add, that amidst the secondary porphyries of the red sandstone, mandelstein and other interposed beds (Germany, Scotland, Hungary) sometimes assume also the aspect of a pyroxenic rock. According to these various relations of position and composition, I think we cannot, in the present state of our knowledge, deny altogether the existence of transition porphyries in the Cordilleras of South America, and to consider all the rocks of syenite, porphyry, and greenstone, which I have just described, as trachytes. The porphyries of the groups §§ 21. and 23. are characterised in South America and Mexico by their constant tendency to a regular stratification; a tendency very rarely observed in Europe, on a great extent of country, in the groups §§ 22. and 24. The regularity of stratification is, however, much greater in the Mexican porphyries posterior to transition clay-slate, than in the porphyries of the Andes of Popayan, Pasto, and Peru, which repose immediately on primitive rocks. I did not observe in this latter formation (§ 21.) one subordinate bed of syenite, greenstone, limestone, and mandelstein, as we find in the groups §§ 22. and 23.

I saw transition porphyries in New Spain, between Acapulco and Tehuilotepec, which were not metalliferous, resting immediately on primitive granite

(Alto de los Caxones Acaguisotla, and several points between Sopilote and Sumpango); but, as farther north, (near Guanaxuato,) metalliferous porphyries of a similar composition cover transition clay-slate, it remains uncertain, notwithstanding the difference of position, if both do not belong to the same formation, and to a more recent formation than the group § 21. A term δ of the geognostic series may immediately follow β, where γ is not developed. Thus Jura limestone, near Lauffenberg, reposes immediately on gneiss, because the intermediary terms of the series of formations, the rocks situated in other places (for instance, in the valley of Necker) between Jura limestone and primitive formations, are suppressed. In the British Islands, according to the observations of the learned professor Buckland, and those of MM. de Buch and Boué, the formation of syenite, greenstone, and transition porphyry, (Ben Nevis, Grampians,) reposes also immediately on primitive rocks (mica-slate and urthon-schiefer). It appears, consequently, to belong to the first group of porphyries, of which I have just traced the history (§ 21.). The porphyries of the north of England and those of Scotland are sometimes covered by grauwacke and sometimes by the coal formation; they have a feldspathic base, and are often found destitute of quartz like the porphyries of equinoxial America. Garnets have been observed in them; this phenomenon is found in the transition porphyries of Zimapan (Mexico), and in those that crown the famous mountain of Potosi, and which probably belong also to the group.

§ 23. If the mandelstein of Hefeld makes a part, as M. de Raumer believes, of the red sandstone formation, the garnetiferous porphyries of Nitzberg (in the Hartz) are probably of secondary formation. In Hungary, garnets occur both in porphyries or porphyritic greenstone of the group § 23., and in the conglomerates of the trachytic formation. It thence results, that the garnets penetrate from the primitive rocks (gneiss, whitestone, serpentine), through transition porphyries, as far as into the trachytes and volcanic basalts, and that in the zones the most distant from each other, certain porphyries present numerous relations with trachytes. I am not informed whether the titaniferous syenite of Keilendorf, in Silesia, which reposes immediately upon gneiss, and passes to a transition granite with small grains destitute of hornblende, belongs to the ancient formation of group § 21., or is a fragment of the formation § 23. placed accidentally on primitive rocks. Nothing is more difficult than to ascertain, with certainty, if there has been a suppression of some intermediary members of the series of rocks, or if the immediate contact which is observed, is that which would be found every where on the globe, in comparing the relative age or position of the same formations.

III. TRANSITION CLAY-SLATE CONTAINING GRAUWACKE, GREENSTONE, BLACK LIMESTONE, SYENITE, AND PORPHYRY.

§ 22. This is the great formation of clay-slate that crosses the western Pyrenees, the Alps of Switzerland between Hartz and Glaris, and the north of Germany from the Hartz as far as to Belgium and Ardennes, and in which grauwacke and limestone predominate; it is the clay-slate and transition gneiss of the Clattentio, Brittany, and Caucasus; it includes the schistose rocks in Norway placed below the porphyries and zircon-syenites, that is, between porphyries and primitive rocks; it is the green clay-slate with black limestone, serpentine, and greenstone of Malpasso, in the Cordillera of Venezuela, and the clay-slate with syenites of Guanaxuato, at Mexico. We have already stated the position of these rocks in the different countries which we have just named; we must now consider them as a whole, and separate the results of geognosy from those ideas which are purely local, and which mineralogical geography presents. The group § 22. reposes, like the two preceding groups, immediately on the primitive rocks; it is distinguished from the former (§ 20.) by the almost total absence of steatitic granular limestone; from the second (§ 21.) by the frequency of clay-slate and grauwackes.

The following formations, closely connected together, belong to this group (§ 22.), which is one of the best known, and most anciently studied: —

Clay-slate, with beds of compact quartz, grauwacke, black limestone, lydian stone, carburetted ampelite, porphyry, greenstone, small-grained granite, syenite, and serpentine.

Grauwacke and quartzose sandstone.

Black limestone.

These rocks are either insulated, or alternate with each other, or form subordinate beds. I have already discussed the characters that usually distinguish primitive from transition clay-slate; I observed that characters drawn from the mineralogical composition of rocks have not the absolute value which has sometimes been assigned to them, and that to employ them with success we must have recourse at the same time to the geognostic situation, the interposition, or absence of fragmentary beds (conglomerates, grauwackes), and the remains of organised bodies, which are totally wanting in primitive formations, and which we begin to find in transition formations. The clay-slates of the latter formation are distinguished by their *variableness*, their continual tendency to change their composition and aspect; by the number of interposed beds; by frequent passages, sometimes rapid, sometimes slow and insensible, to ampelite, siliceous schist, greenstone, or to porphyroid and syenitic rocks. No doubt these changes and effects of internal development are also remarked in some primitive rocks. M. de Charpentier observes, that the gneiss-granites of the Pyrenees, which contain almost always a little hornblende disseminated in the mass, without becoming syenites, and which are believed to be primitive without being more ancient, presents a great number of foreign beds, for instance, beds of mica-slate, greenstone and granular limestone. In the same chain of mountains, primitive mica-slate contains disseminated chiastolite, a substance in general more common in transition clay-slate. The Alps of Switzerland, and principally the passage of Splugen, so well described by M. de Buch, furnishes a mica-slate of primitive formation, which passes insensibly to a porphyry, of which the paste is compact feldspar and contains crystals of lamellar feldspar and quartz. In general, however, these changes are less frequent among primitive than transition formations.

However close may be the connection which we observe between rocks that constitute the same group, or between the different groups of the whole intermediary formation, we see notwithstanding, on different points of the globe, a certain degree of independence, not only between the six groups or terms of the series of transition rocks, (for instance, between clay-slates and grauwacke, porphyries, and syenites,) but also between the partial members of each group or association of intermediary rocks. It thence results, that to understand well those circumstances which characterise the geological constitution of a country, we must study these relations separately (for instance, those of the grauwacke, clay-slate, and limestone, contained in the group § 22., and determine for the different formations or partial members of the same association, their degrees of dependence or independence of each other. We see them either alternating periodically, or enveloping or reducing each other (by an unequal increase of bulk) to the state of simple subordinate beds, or finally covering each other mutually, like primitive rocks of different formations.

It happens in fact, that the partial terms of the same group, α, β, γ, sometimes succeed each other with a certain regularity in a periodical series, α, β, γ, α, β, γ, α. At other times α takes so great a development, that β and γ are found included like simple beds; again, at other times, α, β, γ, are merely superposed, the one on the other, without a periodical return. The latter case does not exclude the possibility that β, before it succeeds to α, may appear at first like a subordinate bed. There happens in one group all that we remark in the non-complex terms of the series of primitive formations. It may be said, as we observed above, that a formation of black limestone, which constitutes great masses of mountains, and is superposed on masses equally considerable of transition clay-slate, indicates, by interposed beds of black limestone, the approach to clay-slate. When β and γ form included beds in α, those beds may be so frequently repeated, that on a great extent of space they may assume the aspect of alternating rocks. Thus intermediary clay-slate, which at first enveloped grauwacke and black limestone, and then alternated with them, (defile of Aston in the Pyrenees, Maxen in Saxony,) finishes by covering and with a great increase of mass, these alternating rocks, or frequently interposed beds. The regularity of the type in the partial formations of every group, is like the direction of strata on the angle made by those strata with the meridian. Every thing appears at first sight confused and contradictory; but when all is examined with care, on a great extent of country, we constantly end by recognising certain laws of position or stratification. If the type which we discover in the suite of partial formations appears to vary according to localities, it is because the development of those small formations has not been every where the same. Sometimes (Caucasus) porphyry, limestone, syenite, and transition granite are found developed at the same time in transition clay-slate; at other times it contains neither porphyry (Cotentin, Alps of Switzerland) nor grauwacke (chain on the shore of Venezuela), nor transition granite and syenite (Pyrenees). The association of transition clay-slate and compact black limestone is almost as constant as that of white granular limestone with mica-slate in the primitive series. We find, however, transition limestone, which being associated neither to clay-slate, or grauwacke, appears to replace clay-slate geognostically; but I know not a single point on the two continents, where transition clay-slate has been seen on a space at all considerable, without being accompanied by limestone.

We have just seen that in some parts of the globe (Caucasus and the peninsula of the Cotentin) intermediary clay-slate contains either porphyries, or syenites and granites; in other parts (Norway and Saxony, between Friedrichswalde, Maxen and Dohna) these three rocks, after having appeared as subordinate beds to clay-slate, are found placed over it, either insulately and forming considerable masses, or alternating together. It is only in these cases of separation or alternation, that an *independent formation of porphyry* (Mexico), or an *independent formation of porphyry and syenite* (Norway), seem to cover the formation of intermediary clay-slate. The same separation (if not the same independence) is sometimes observed in transition limestone; and, although in a less striking degree, in the grauwackes.

Syenite and granite are connected in the transition series rather with porphyries than with mica-slate or gneiss; we find, in the same series, syenite without granite; but it is much more rare to find syenite and granite without porphyry. When the partial members of a group, α, β, γ, alternate in periodical succession, and consequently are not interstratified the one with the other like subordinate beds, nor superposed like distinct rocks or formations, it is difficult to determine if β and γ are a more recent formation than α; even, however, in the case of what is called contemporary origin, we recognise, by an attentive examination, certain *preponderances* of formation. Grauwacke and transition clay-slate are in general more ancient than black limestone; or, to support my opinion by a very just observation of M. de Charpentier, "it is generally observed that, notwithstanding the alternation in that part of the intermediary formation which is nearest to the primitive formation, it is grauwacke and clay-slate that predominate in great masses, and to which limestone is subordinate; while, on the contrary, in the newest part of the transition formation, limestone is the preponderating rock, and clay-slate is only found subordinate to limestone in beds more or less thick."

After having stated the relations of age and position of rocks which constitute the same group, we shall now more particularly characterise each of the partial formations.

Clay-slate, blackish-blue, and carburetted, or greenish, unctuous, and of a silky lustre, sometimes earthy, and in very thick layers, sometimes in very thin laminæ. In the very ancient beds that pass to transition mica-slate it is undulated, and presents only large scales of mica strongly adherent. In newer beds, near the contact with grauwacke, it contains very small insulated scales of mica, and frequently also chiastolite, epidote, and veins of quartz. Transition clay-slate, characterised by its extreme *variableness*, that is, by its continual tendency to change its composition and aspect, contains a great number of beds, some of which, by their frequent repetition, appear to form with it alternating rocks. The most usual effects of this internal development are the interposed beds of *grauwacke* and grauwacke-slate; of *limestone* generally black and compact, or dark grey, sometimes reddish (Braunsdorf), and even granular and white (Miltitz in Saxony), as in the group §20.; of *greenstone-porphyry* (Caucasus, Saxony, near Friedrickswalde and Seidwitzgrund); of *alum-slate*, or ampelite highly carburetted; of *compact quartz* (quartzite, quartzfels of Hausmann), sometimes with small crystals of feldspar (Kemielf in Finland); *lydian stone* and siliceous schist. These two latter siliceous substances are found in the formations of clay-slate, grauwacke, limestone, and in the form of jasper in porphyry; they prove by their presence the geognostic affinity that exists between these various transition rocks. Clay-slate (§ 22.) contains less usually subordinate beds of *gneiss* (Lokwitzgrund and Neutanneberg); *mica-slate* and *granite* (Krotte in Saxony), Furstenstein in Silesia, Honfleur in Normandy, Montherme in the Ardennes; *granite* and *syenite* (Caucasus, Cotentin, Calixelf in Norway); *graphic schistose clay* (schwarze kreide; valley of Castillon in the Pyrenees, Ludwigstadt in Franconia); *whet-slate* (wetz-schiefer); *serpentine* (Bochetta, near Genoa, Lovezara and two other points further north, towards Voltaggio; see § 19.); *com-*

pact feldspar (valley of Arran, in the Pyrenees, Poullaouen in Bretagne), sometimes pure, dark greenish-grey or olive-green, sometimes (Pyrenees, Hartz, and the eastern part of Upper Egypt) mixed with disseminated crystals of lamellar feldspar, hornblende, schorl, and quartz. When compact feldspar is simply mixed with hornblende, it forms the *grunstein-schiefer* of Werner, which alternates with transition clay-slate (Allenborg in Sweden), and is found in primitive formations. Although, as I endeavoured to prove in my Memoir on the βασανιτης and λιθος 'Ηρακλεια, published in 1790, the major part of the basalts of the ancients belong to syenitic transition rocks, or to beds of greenstone contained in primitive rocks, yet the examination of the Egyptian statues, preserved at Rome, Naples, London, and Paris, gave rise in my mind to the idea that many of the *black* and *green basalts* of our antiquaries are only masses of compact feldspar, from intermediary formations, and coloured either black or green by hornblende, chlorite, carbon, or metallic oxids. Nothing but a chemical analysis of these ancient unmixed masses could solve this question of mineralogical archaiology. M. Beudant saw in the transition formation of Hungary, porphyroid greenstone passing into a green or black paste of homogeneous appearance. This paste was only compact feldspar coloured by hornblende.

We have already observed above, that transition clay-slate forms far greater masses in the globe than primitive clay-slate. The latter is generally subordinate to mica-slate; as an independent formation, it is as rare in the Pyrenees and the Alps, as in the Cordilleras. In South America, between the parallels of 10° north, and 7° south, I saw transition clay-slate only on the southern declivity of the littoral chain of Venezuela, at the entrance of the Llanos of Calaboza. The basin of the Llanos, the bottom of an ancient lake covered with secondary formations (red sandstone, zechstein, and clay-gypsum), is bounded by a band of intermediary formation of clay-slate, black limestone, and euphotide, connected with transition greenstone. Gneiss and mica-slate, between the valleys of Avague and the Villa de Ceura, constitute only one formation on which clay-slate reposes in conformable position, in the ravines of Malpasso and Piedras azules (direction N. 52° E.; inclin. 70° towards the N.W), of which the lower beds are green, steatitic, and mixed with hornblende, and the upper are of a greyish-green, and darkish-blue colour. This clay-slate contains (like that of Steben, in Franconia, the duchy of Nassau, and of Peschels-Mühle in Saxony) beds of greenstone, sometimes massive, at other times of a globular structure.

The famous vein of Guanaxuato in New Spain, which, from 1786 to 1803, has produced yearly, on an average, 556,000 marks of silver, also traverses transition clay-slate. This rock, in its lower strata, passes, in the mine of Valenciana (at the height of 932 toises above the level of the sea,) to a talcose slate; and I described it in my *Political Essay*, as placed on the limit of the primitive and intermediary formations. A more particular examination of the relations of position which I noted on the spot, and the comparison of the beds of syenite and serpentine which have been pierced in digging the *general tiro*, with the beds that are interposed in the transition formations of Saxony, of the Bochetta of Genoa, and of the Cotentin, convinced me that the clay-slate of Guanaxuato belongs to the most ancient intermediary formations. We do not know whether its stratification be *parallel* and *conformable* with that of the granite-gneiss of Zacatecas, and of Penon blanco, which probably supports it, the contact of these formations not having been observed; but almost all the porphyry-rocks on the great table-land of Mexico follow the general direction of the chain of mountains. (N. 40°, 50° W.) This *perfect concordance* (Gleichfurmigbeit der Lagerung) has been observed between the primitive gneiss and the transition clay-slate of Saxony (Friedrickswalde; valleys of Müglitz, Seidewitz, and Lockwitz); it proves that the formation of the intermediary series immediately succeeded that of the last beds of the primitive series. In the Pyrenees, as M. de Charpentier has observed, the former of these series is found in a different position (not parallel), sometimes in *transgressive position* (Ibergreifende Lagerung) with the latter. I shall observe, on this occasion, that the parallelism between the stratification of two *consecutive formations*, or the absence of this parallelism does not alone decide the question whether the two formations are united or not in the same primitive or secondary series; it is rather the sum of these geognostic relations that solves that problem. The clay-slate of Guanaxuato is very regularly stratified (direct. N. 46° W.; inclin. 45° S.W.), and the form of the valleys has no influence on the direction or inclination of the strata. We distinguish there three varieties, which may be considered as three epochas of formation; a clay-slate of a silvery lustre, and steatitic, passing to a talc-slate (talk-schiefer); a greenish clay-slate, with a silky lustre, resembling chlorite slate; finally, a black clay-slate, in very thin laminæ, surcharged with carbon, staining the figures like ampelite and the marly slate of zechstein, but not effervescing with acids. I have named these varieties in the order in which I observed them from below upwards, in the mine of Valenciana, which is 263 toises in perpendicular depth; but in the mines of Mellado, Animas, and Rayas, the surcarburetted clay-slate (*hoja de libro*) occurs beneath the green and steatitic variety; and it is probable, that the strata which pass to talc-slate, chlorite, and ampelite, alternate several times with each other.

The thickness of this formation of transition clay-slate, which I found at the mountain of Santa-Rosa near Los Joares, where the Indians collect ice in small artificial basins, is more than 3000 feet. It contains in subordinate beds, not only syenite (like the transition clay-slate of the Cotentin), but also, which is very remarkable, serpentine, and a hornblende-slate that is not greenstone. In digging in the massive rock, the *tiro general de Valenciana*, which has cost nearly seven millions of francs, the following strata were found, reckoning downwards for ninety-four toises of depth: an ancient conglomerate, representing the red sandstone; black transition clay-slate, strongly carburetted, in very thin laminæ; clay-slate, bluish-grey, and containing magnesia; hornblende-slate, greenish-black, a little mixed with quartz and pyrites, destitute of feldspar, not passing to greenstone, and altogether similar to the hornblende-slate (hornblend-schiefer) which forms beds in primitive gneiss and mica-slate (§§ 5. and 11.); green serpentine of uneven fracture and fine grain, dull, but translucent on the edges, containing much pyrites, destitute of garnets and diallage metalloid (schillerspath), mixed with talc and steatite; hornblende-slate; syenite, or a granular mixture of much darkish-green hornblende, yellowish quartz, and a little lamellar and white feldspar. This syenite splits into very thin layers; the quartz and feldspar are so irregularly spread, that they sometimes form small veins in a paste of hornblende. The syenite is the largest of these eight interposed beds, of which the direction and inclination are exactly parallel to that of the whole rock; it is more than 30 toises thick; and, as I saw in the deepest working of the mine (*planes* of San Bernado), at 172 toises below the bed of syenite, carburetted clay-slate occurring again, identical with that in which new shafts are beginning to be sunk, no doubt can remain that hornblende-slate alternating twice with serpentine, and serpentine alternating probably with syenite, form beds subordinate to the great mass of clay-slate of Guanaxuato. The connection which we have just remarked between hornblende-rocks and serpentine, is found in other parts of the globe, in formations of euphotide of different ages; for instance, at Heidelberg near Zell in Franconia (§ 19); at Keilwig in the northern extremity of Norway; at Portsoy in Scotland, and and at the island of Cuba, between Regla and Guanavacoa.

I saw no remains of organised bodies, nor beds of porphyry, grauwacke, nor lydian stone, in the transition clay-slate of Guanaxuato, which is the rock richest in silver that has hitherto been found; but this clay-slate is covered in conformable position in some places by transition porphyries very regularly stratified (los Alamos de la Sierra); in some by greenstone and syenites alternating thousands of times together (between Esperanca and Comangillas; and in others either by a calcareous conglomerate or by transition limestone of a bluish-grey colour, mixed with clay, and fine-grained (ravine of Acabuca), or by red sandstone (Marfil). These relations of the clay-slate of Guanaxuato, with the rocks which it supports, and some of which (the syenites) appear first as subordinate beds, suffice to place it among the transition formations; above all, they justify this result in the opinion of those geognosts who are acquainted with the observations which have been recently published on the intermediary formations of Europe. With respect to lydian stone, there can be no doubt that it is contained on some points not yet explored, in the clay-slate of Guanaxuato; for I found the former substance frequently imbedded in large fragments in the ancient conglomerate (red sandstone) which covers the clay-slate between Valenciana, Marfil, and Cuevas. Ten leagues to the south of Cuevas, between Queretaro and la Cuesta de la Noria, in the middle of a Mexican table-land, a transition clay-slate appears beneath the porphyry, darkish-grey, and passing both to siliceous slate (schistoïd jasper kiesel-schiefer) and lydian stone. Many fragments of this latter substance are found near the Noria, scattered in the fields. The rocks with argentiferous veins of Zacatecas, and a small part of the veins of Catorce, according to the report of two well informed mineralogists, MM. Sonneschmidt and Valencia, also traverse transition clay-slate, which contains true beds of lydian stone, and which appears to rest on syenites. This superposition would prove, according to what has been observed of the *tiro general of Valenciana*, that the clay-slates of Mexico constitute (as at Caucasus and in the Cotentin) the same formation with the transition syenites and euphotides, and that perhaps they alternate with the latter rocks.

Grauwacke. This barbarous name, employed by

German and English geognosts, has been preserved, like that of thonschiefer (clay-slate), to avoid that confusion of names, so prejudicial to the study of formations. It designates, when taken in a more general sense, every conglomerate, sandstone, pudding-stone, fragmentary or arenaceous rock of transition formation, that is anterior to the red sandstone and coal formation. The *old red sandstone* of Herefordshire (of Mr. Buckland), placed beneath the mountain-limestone (transition limestone of Derbyshire), is a sandstone of the intermediary series, as that excellent geognost has himself indicated in his memoir on the structure of the Alps. The *new red conglomerate* of Exeter is the red sandstone of the French mineralogists, or *todte liegende* of the Germans; it is the first sandstone of the secondary series, that is, the sandstone of the coal formation, which is closely connected with the secondary porphyry, which is called on that account, the porphyry of the red sandstone. When the word *grauwacke* (*tromate* of M. d'Aubuisson, ancient psammite, and quartzose mimophyre of M. Brongniart) is used in a more confined sense, it is applied to the arenaceous transition rocks, which contain only small fragments of simple substances, more or less rounded, for instance, of quartz, of lydian stone, of feldspar, and of clay-slate, but not fragments of composed rocks. The grauwackes are then excluded, and we give the name of breccias or *conglomerates, with large primitive fragments* (§ 20.), to the various agglutinations of pieces of granite, gneiss, and syenite; calcareous pudding-stones are also separated, in which rounded fragments of carbonate of lime are cemented by a paste of the same nature. All those distinctions (if we except some calcareous breccias, in which the contained and the containing matter may very probably be sometimes of contemporary origin) are of no great importance to the study of formations. Coarse grauwacke (grosskornige grauwacke) passes by degrees to a conglomerate with large fragments; it alternates in the same country, not only with beds of fine-grained grauwacke but also with others the paste of which is almost homogeneous. The pudding-stones and breccias with large fragments of primitive and composed rocks (urfels-conglomerate of Valorsine in Savoy, and of Salvan in the Bas Valais), are true grauwackes; they are the most ancient beds of this formation, in which fragments with distinct outlines are not blended into the mass, and of which the slaty cement, with curved and undulated laminæ, resembles mica-slate, while the cement of the more recent grauwackes of the Hartz, the duchy of Nassau, and Mexico, resembles clay-slate. In general, the conglomerates, or grauwackes of the group § 20. present fragments of pre-existing rocks of a much more considerable and unequal bulk than the grauwackes of the group § 22.

When we compare these with transition limestone, we find them most frequently of anterior origin; they sometimes even replace transition clay-slate. The priority of grauwacke to the limestone is evident in the Pyrenees, and in Hungary. It appears that in the latter country transition clay-slate has not assumed a great development; since, far from being there an independent formation containing grauwacke, it is, on the contrary, grauwacke-slate (grauwacken-schiefer) with agglutinated scales of mica, which here assumes all the characters of a real transition slate. In England also, the great insulated mass of *mountain-limestone* (counties of Derby, Gloucester, and Somerset) is of a later date than the great mass of grauwacke that alternates with limestone strata; but when we examine in detail the points where the different members of the group § 22. have asumed an extraordinary development, we recognise two great limestone formations (the transition limestone of Longhope, and the mountain-limestone of Derbyshire and of South Wales) alternating with two formations of grauwacke (grauwacke of May-Hill, and the old red sandstone of Mitchel-Dean in Herefordshire). This order of position, this bisection of calcareous and arenaceous masses, is found repeated in several parts of the globe. M. Beudant has recognised, in Hungary, the *old red sandstone* of England in the quartzose transition limestone of Neusold, which lies upon coarse-grained grauwacke, after having been interposed in it; he thinks he recognises the mountain-limestone, placed between the *old red sandstone*, and the coal formation of England, in the intermediary limestone of the group of Tetra. If the Oldenhorn and the Diablerets belong, which is very probable, to the transition series, there is also in Switzerland, above and below the grauwacke of the Dent de Chamossaire, two great formations of black limestone, which M. de Buch has long since distinguished by the names of first and second transition limestone. In Norway (Christianifiord), the grauwacke is certainly newer than the intermediary clay-slate, and the orthoceratite-limestone.

In the centre of Europe, very fine-grained grauwacke sometimes contains fragments of crystals of lamellar feldspar, which give it a porphyroid aspect (Pont Pelessier, near Seroox; Elm, in the passage of Splugen, Nevsohl, in Hungary); but we must not confound these varieties of an arenaceous rock with beds of interposed porphyry. We shall have occasion to remark, that, in both continents, these broken crystals of feldspar are found in red sandstone, and in a feldspathic conglomerate much more recent. In the southern hemisphere grauwacke forms, according to M. Eschwege, the eastern declivity of the mountains of Brazil. I found this same rock in the United States (chain of the Alleghanys) containing beds of lydian stone and black limestone, exactly similar to those of the transition formation of the Hartz. M. Maclure first determined the real limits of grauwacke, from Carolina as far as Lake Champlain. In the north of England (Cumberland, Westmoreland,) this formation furnishes beds of garnetiferous porphyry.

Transition limestone. This rock commences, either by forming beds in grauwacke and intermediary clay-slate, or by alternating with them; the clay-slate and grauwacke-slate afterwards disappear, and the superposed limestone becomes a *simple formation,* which we should be tempted to call independent, although it still belongs to the group § 22. When there is an alternation of schist and limestone, that alternation takes place either by thick beds (summit of the Bochetta near Genoa, and the road between Novi and Gavi) as in the *composed formations* of granite and gneiss, grauwacke and grauwacke-slate, syenite and greenstone, clay-slate and porphyry; or, the alternation extends to the thinnest layers of the rocks (calschistes), so that every layer of schist is cemented to one of limestone (valleys of Campua and Oneil, in the Pyrenees, and mountains of Ponik in Hungary).

In the same manner as we find in the Pyrenees, interposed in granite-gneiss and in primitive mica-slate, beds of limestone, which, from their aspect only, might be considered as intermediary, viz. greyish-black limestone (Col de la Trappe) coloured by graphite which is the most ancient of the carbonaceous substances, fetid limestone giving the smell of sulphuretted hydrogen, and compact limestone filled with chiastolite; so do the transition formations also of the group § 22. exhibit some examples of white and granular limestone,

(Miltitz in Saxony, valleys of Ossan and Soubic in the Pyrenees). In general, however, if we except the group § 20. (that of which the Tarentaise affords the type), the limestones of intermediary formation are either compact, or pass to fine-grained granular limestone. Their colours are darker (raven-grey, dark-grey,) than those of primitive limestones. The greatest number of the fine varieties of red, green, and yellow marble), (valley of Luchon in the Pyrenees,) celebrated among antiquaries by the names of *African flowered marble, black of Lucullus, antique yellow and red, pavonazzo and gilded breccia,* appear to me to belong to the limestones and calcareous conglomerates of transition. We have seen above, that the chiastolite of transition clay-slate appears as an exception in primitive clay-slate; in a similar manner, tremolite, so common in dolerite and primitive white limestone, occurs (between Giellebeck and Doamea in Norway) in black transition limestone. Some mineral species, no doubt, belong more to one particular age than to any other; but their relations with the formations are not sufficiently exclusive to form diagnostic characters in a science in which the position only ought decide positively. Local circumstances have often a singular influence on the connection between mineral species and formations. In the Pyrenees, and chiefly in South America, disseminated garnets are peculiar to gneiss, whilst every where else they seem rather to belong to mica-slate.

England and Scotland, some traces of coal that differs from anthracite.

The true *variolites* (Durance, Mont Rose) which contain nodules of compact feldspar disseminated in an intimate and almost homogeneous mixture of hornblende, chlorite, and feldspar, belong either to the group we have just described, or to the following group. They are perhaps only subordinate beds in greenstone-porphyry, beds in which a certain quantity of feldspar has separated from the mass. These variolites were long thought to be only pebbles, or large detached fragments; they must not be confounded with the variolites with nodules of calc-spar (blatterstein) subordinate to green transition clay-slate, nor with the variolites that are occasioned by infiltration into the mandelstein of the red sandstone.

Although we are yet far from being able to complete the history of every intermediary and secondary formation, by the enumeration of the species of fossil bodies that are found in them, we shall, however, indicate some of those organic remains which seem to characterise the group § 22. In *clay-slate* and *grauwacke*, and principally in grauwacke-slate; monocotyledon plants (arundinacées or bambousacées) perhaps anterior to the most ancient animals, entrochites, coralites, ammonites (valleys of the Castillon in the Pyrenees; base of the mountain of Fis, in Savoy; duchy of Nassau, and the Hartz in Germany); hysterolites, orthoceratites, much more rare than in intermediary limestone, pectinites (Gerolstein in Germany); trilobites of M. Wahlenburg, in which we discern no trace of eyes (Olstorp in Sweden); ogygies of M. Brongniart, in which the eyes may be said to be indicated only by two tuberosities on the scutum (Angers and North America); calymene of Tristan, and calymene-macrophtalme of Brongniart (Bretagne, Cotentin). In the limestone, viz. in the most ancient beds, entrochites, madrepores, belemnites (Bex, in Switzerland; Peak of Bedillac in the Pyrenees); sometimes ammonites, never in beds, but insulated; orthoceratites, Asaphus Buchii, A. Hausmanni (Wales, Sweden); very few bivalve shells. In the newer beds of limestone, Calymene Blumenbachii (Dudley in England, and Miami in North America); Asaphus caudatus of Brongniart, ammonites, terebratulites, orthoceratites, some gryphites (Namur, Avesnes); and encrinites. In Germany, the transition limestone is sometimes (Eiffel, and duchy of Bergen) entirely filled with shells. The granular limestone of the Isle of Paros (Link, *Urwelt*, page 2.) contains organic remains, according to a passage of Xénophane of Colophon, preserved in Origen (*Philosophumena*, c. xiv., t. i., p. 893., B. *edit. Delarne*); but it is still doubtful, from what we read, δαφνη or αφυη, if those remains are of the vegetable kingdom (the wood of the laurel), or of the animal kingdom (impression of an anchovy). We do not insist on this observation, for it is possible that the marble of Paros may be as little primitive as that of Carara, on which I have the same doubts as several celebrated geognosts. The phenomenon of caverns does not oppose, however, the high antiquity assigned to the limestones of the Archipelago. In some countries (Silesia, near Kaufungen; Pyrenees, valleys of Naupounts, and mountain of Meigut) there are caverns in rocks that appear to be primitive limestone.

IV. & V. PORPHYRIES, SYENITES, AND GREENSTONE POSTERIOR TO TRANSITION CLAY-SLATE, SOMETIMES EVEN TO LIMESTONE WITH ORTHOCERATITES.

§ 25. I place in two groups, what perhaps form but one, the porphyries, greenstone-porphyries, and syenites, which I have seen in both hemispheres covering transition clay-slate. These rocks, by their composition and their relations with trachytes placed immediately on them, present a great analogy with the more ancient group § 21. It is in these porphyries and porphyritic greenstones, that, at the north of the equator, in Mexico and in Hungary, the immensely rich gold and silver mines have been discovered; for, although the metalliferous rock of Schemnitz (*saxum metalliferum* of Born) may perhaps be posterior to the transition limestones, containing some indistinct organic remains, this position, in the opinion of a celebrated geognost, M. Beudant, is too uncertain to separate formations so closely allied as those of New Spain and Hungary. The syenites with zircons, the transition granites and porphyries of Norway, which MM. de Buch and Hausmann have made us acquainted with, are not only posterior (Stromsoë, Krogskoven) to grauwacke and a clay-slate that alternates with orthoceratite-limestone, but these rocks also cover (Skeen) immediately a quartzite (quartzfels) that represents grauwacke, and reposes on a black limestone destitute of alternating beds of clay-slate.

From these considerations it results, that there are sufficient reasons for uniting the groups §§ 23. and 24., distinguishing among transition porphyries only two independent formations, one anterior, the other posterior to clay-slate, and a third formation (§ 22.) subordinate to that rock. The property which certain porphyries and porphyritic syenites possess of being eminently metalliferous, ought not, I think, to oppose the union of the rocks of Mexico, Hungary, Saxony, and Norway. The ores of gold and silver do not form contemporary beds, but are veins of extraordinary size. Some transition porphyries, several of which we should be tempted to place among trachytes because they contain true beds of phonolite with glassy feldspar, participate in these mineral riches, which, among the rocks posterior to primitive formations, were long thought to be found exclusively in carburetted and micaceous clay-slate, grauwacke, and transition limestone. There exist in the same regions groups of porphyries and syenites, very analogous in their mineralogical composition and their position, to the rocks containing the rich mines of Schemnitz or New Spain, and which, nevertheless, are found entirely destitute of metals. This is the case with almost all the transition porphyries of South America. The great workings of Peru, those of Hualgayoc or Chota, and Llauricocha or Pasco, are not in porphyry, but alpine limestone. The famous Cerro del Potosi, in the republic of Buenos-Ayres, is composed of clay-slate (transition ?) covered by porphyries that contain disseminated garnets.

If the great argentiferous and auriferous deposits that have formed for ages the wealth of Hungary and Transylvania, are found solely amidst syenites and porphyritic greenstones, we must not thence conclude that it is the same in New Spain. The Mexican porphyries no doubt offer insulated examples of immense riches. At Pachuca, the only pit of del Encino furnished alone annually, during a long time, more than 30,000 marks of silver; in 1726 and 1727, the two workings of la Biscaina and Xacal gave together 542,000 marks, that is, almost twice as much as all Europe and Asiatic Russia produced in the same interval. These same porphyries of Real del Monte, which are connected by their upper beds with porphyritic trachytes and pearlstone, with obsidian of Cerro de las Navajas, furnished by the working of the mine of la Biscaina, to the count of Regla, from 1762 to 1781, more than eleven millions of piasters. These riches, however, are still inferior to those which are drawn in the same country from transition formations which are not porphyritic. The veta negra of Sombrerete, which traverses a compact limestone containing nodules of lydian stone, has furnished the example of the greatest abundance of silver which has been observed in the two worlds; the family of Fagoaga, or of the Marquis del Apartado, drew from thence in a few months, a neat profit of four millions of piasters. The produce of the mine of Valenciana, worked in transition slate, has been so constant, that to the end of the last century, it never ceased to furnish annually, during forty years successively, above 360,000 marks of silver. In general, in the central part of New Spain, where porphyries are frequent, it is not that rock which affords the precious metals in the three great workings of Guanaxuato, Zacatecas, and Catorce. These three mining districts, which yield the half of all the Mexican gold and silver, are situated between the 18° and 23° of north latitude. The miners there work on metalliferous mineral deposits, almost entirely in intermediary formations of clay-slate, grauwacke, and alpine limestone; I say almost entirely, for the famous *Veta madne* de Guanaxuato, richer than Potosi, and furnishing till 1804, on an average, a sixth of the silver which America pours into the circulation of the whole world, traverses both clay-slate and porphyry. The mines of Belgrado, San Bruno, and Marisanchez, opened in the porphyritic part at the south-east of Valenciana, are but of small importance. Other workings carried on the porphyries of the group § 23. (Real del Monte, Moran, Pachuca, and Bolaños), do not now furnish above 100,000 marks, or a twenty-fifth part of the silver exported (1803) from the port of Vera Cruz. I thought it was here proper to state these facts, because the denomination which I have often used in my works of *metalliferous porphyries*, might lead to the error of considering the metallic riches of the new world as procured in great part from transition porphyries. The more we advance in the study of the constitution of the globe in different climates, the more we are convinced that there scarcely exists one rock anterior to alpine limestone which has not been found in some countries extremely argentiferous. The phenomenon of these ancient veins in which our metallic riches are deposited (perhaps as the specular iron and muriate of copper are deposited in modern times in the fissures of lava) is a phenomenon that appears is some degree independent of the specific nature of rocks.

To give a precise idea of the composition of the porphyry, syenite, and greenstone formation, posterior to transition clay-slate, it is necessary, in the present state of the science, to distinguish four *partial formations*, viz.—those

of the equinoxial region of the new continent,
of Hungary,
of Saxony, and
of Norway.

Notwithstanding the relations that conceal these partial formations, each of them exhibits very remarkable differences. We shall designate them by simple geographical names, according to the places which supply the most distinct *types*, without inferring from thence that the formation of Hungary cannot be found in the new continent, or that of Guanaxuato, with all the circumstances that accompany it, in some parts of Europe.

A. *Groups of the equinoxial region of the new continent.*

a. *In the northern hemisphere.* What in general characterises the porphyries, in part extremely metalliferous, of equinoxial America (those of group § 23. as well as those of group § 21.), is the almost total absence of quartz, and the presence of hornblende, glassy feldspar, and sometimes of pyroxene; I have insisted on these distinctive characters in all the works which I have published since 1805; they are also found in great part in the porphyries or porphyritic greenstones, equally metalliferous, of Hungary and Transylvania. The Mexican porphyries, as we have observed above, often contain two varieties of feldspar, the common and the glassy; the former resists decomposition much less than the latter. They are almost as well distinguished by the form of their crystals, whether large or slender, as by the lustre and lamellar structure, more or less distinct. The quartz, if it sometimes appears, is not crystallised, but in small irregular grains; pyroxene and garnet, which are found also in the porphyritic greenstone of Hungary, are very rare.

The argentiferous Mexican group abounds less in hornblende; mica, which is found in some trachytes, is always wanting in the porphyries of New Spain. These rocks are for the most part very regularly stratified; and also the direction of their strata is often (between Moxonera and Sopilote, at the north of Acapulco, at Puerto de Santa Rosa, near Guanaxuato,) concordant with the direction of the primitive and intermediary rocks to which they are superposed. In New Spain, as well as in Hungary, the trachytic formation is placed immediately on metalliferous porphyries; but in the former, the porphyries are covered on some points (Zimapan, Xaschi, and Xacala) by darkish-grey transition limestone, on some (Villalpando) by red sandstone, and on others (between Masatlan and Chilpanzingo; between Amajaque and Magdalena; between San Francisco Ocotlan and la Puebla de los Angeles; between Cholula and Totomehuacan) by alpine limestone.

The transition porphyries of Hungary, Saxony, and Norway, are of a very complicated structure; they alternate with syenites, granites, and greenstones; and when there is no *alternation*, these three latter rocks, and even the mica-slate and steatitic limestone are found contained as subordinate beds, in porphyries. The abundance of those subordinate beds separates, in a very decided manner, the porphyries of Hungary and Norway from the trachytic rocks; it also removes from the porphyries of New Spain, which resemble them in their mineralogical composition (by the nature of their paste, and imbedded crystals). There is a great simplicity in the structure of the Mexican porphyries; they form an immense and uninterrupted series of subordinate beds. I saw syenites in the transition clay-slate of Guanaxuato (§ 20.); I have seen them above this clay-slate, alternating with greenstone; but I have not found either syenite, or mica-slate, or greenstone, or limestone, in the porphyries of Moxonera, of Pachuca, of Moran, and of Guanaxuato. It is only at Bolaños that mandelstein is found in porphyry. This uniform and uninterrupted development of the metalliferous and non-metalliferous porphyries of New Spain is a very striking phenomenon; it renders the systematic separation of porphyry and trachyte formations more difficult, where those rocks immediately support each other. When we estimate the thickness of the two united formations, that is, when we ascend from the lowest beds of porphyry which we conceive to be transition because it is covered with great formations of limestone analogous to zechstein, (Guasintlan, at the western declivity, and Venta del Encero, at the eastern declivity of the Cordilleras,) as far as the trachytic summit of the great volcano, de la Puebla (Popocatepetl), we find, according to my barometric and trigonometric measures, a thickness, uninterrupted by interposed rocks, of more than 13,000 French feet (2233 toises). The thickness of the beds of metalliferous porphyry only, in reckoning from Guasintlan, and Puente de Istla (where the porphyries are hid beneath the porous mandelstein of Guchilaque, and the valley of Mexico), as far as the upper part of the argentiferous veins of Cabrera (Real de Moran), is 5000 feet (807 toises). These dimensions were determined by comparing the absolute heights of the stations; for, according to the variable inclination of the beds, and the relation between the direction of the sections, and that of the rock, it is probable that the *apparent thickness* (the difference between the *maximum* and the *minimum* of heights) is very little remote from the *real thickness*, which is the sum of the thickness estimated perpendicularly at the fissures of stratification. The following are the most interesting local circumstances respecting the position of the porphyries of Mexico, between the 17° and 21° of north latitude.

Road from Acapulco to Mexico. The porphyry at the western declivity of the Cordillera of Anahuac, descends only as far as the valley of Rio Papagalla, a little to the north of the Venta de Tierra Colorada, at the height of 230 toises above the level of the Pacific Ocean. On the eastern declivity of the Cordillera of Anahuac, between the valley of Mexico and the foot of Vera Cruz, I saw no traces of this rock below the Encero, at the height of 476 toises. The porphyries are hid beneath an argillaceous sandstone, in which fragments of amygdaloidal trachyte are imbedded. The two principal groups of porphyry, in the road from

Acapulco to Mexico, are those of Moxonera and Zumpango.

The granitic valley of Papagallo is bounded at the south (Alto del Peregrino) by a formation of compact limestone (85 toises thick), darkish-blue, and traversed by small white veins of calcareous spar. It is full of great caverns, but it is rather analogous to the alpine limestone, than the transition limestone. The valley is bounded to the north by a mass of porphyry (Alto de la Moxonera and de los Caxones) 355 toises thick. This porphyry is rather regularly stratified (dir. N. 35° E.; inclin. 40° at N.W.); sometimes it is divided in balls with concentric layers. Its base is greenish and argillaceous, containing glassy feldspar imbedded, and decomposed pyroxenes, which have almost the colour of olivine; but no quartz, mica, nor lamellar feldspar. Great masses of reddish-white clay are interposed in this porphyry, which rests immediately, like the limestone of Peregrino, (of which the direction of the strata is N. 45° E.; inclin. 60° at N.W.) on primitive granite. The latter, which has been described above (§ 7.), contains at the foot of the porphyritic hill of Los Caxones, even in the valley of Papagallo, veins of black amphibolite, and balls of granite with concentric layers, similar to those which I observed at Fichtelgebirge, near Seissen. The greatest mass of this coarse-grained granite is very regularly stratified (dir. N. 40° E.), and inclined by groups of vast extent, most frequently at N.W., sometimes at S.E. The neighbouring summits (porphyritic? Cerros de las Caxas, and del Toro,) are of singular forms; and, if from the mineralogical composition of the porphyry of Moxonera and the Alto de los Caxones, and from its insulated position, we should be led to consider it as trachyte, the parallelism of the direction of these strata with those of the limestone and granite, and the covering by a very similar neighbouring porphyry (Masatlan,) might be opposed to that hypothesis. In descending the porphyritic mountain of Los Caxones towards the south, that is, towards the coasts of the Pacific ocean, I saw, alternately appearing, the primitive granite of the valley of Papagallo, the alpine limestone of the Alto del Peregrino, the primitive granite of the valley of Camaron, the syenite of the Alto del Camaron, and finally, the primitive granite of Exido and the coast of Acapulco. The syenite of Camaron, containing crystals of hornblende about three-quarters of an inch long, does not appear to me to be allied to the Mexican porphyries. It is but a change in the composition of the mass of granite, which, in this region, is mixed with hornblende, and becomes porphyritic on all the summits of the hills.

The second group of intermediary porphyries of which I could carefully examine the superposition, is that of Zumpango. This group begins some leagues to the north of the Alto de los Caxones, and in stretching towards Mescala, supports a vast table-land, composed of limestone, sandstone, and gypsum (between Masatlan and Chilpansingo). In this table-land, the absolute height of which is 700 toises above the level of the sea, a porphyry, similar by its composition to that of Moxonera, supports secondary formations of a very complicated structure. In descending the Alto de los Caxones (height 585 toises) towards the north, we at first see the primitive granite of the valley of Papagallo reappear; we then discover an alpine limestone, similar to that of Peregrino (200 toises broad, which is immediately superposed on granite); the granite then appears again; and finally, we reach the porphyritic group of Zumpango, in which the direction of the strata is very regularly preserved, N. 30°, at 45° E., with a frequent inclination to N.W.

This porphyry, containing glassy feldspar, destitute of hornblende, and covering primitive granite, serves first for a basis (Acaguisotla) to a formation of amygdaloid, that is reddish-brown, semi-vitreous, almost without cavities, containing nodules of decomposed calcedony, and scales of black mica and melanite. The mandelstein soon disappears, and porphyry again occurs of very considerable extent, till it is hid beneath the limestone of Masatlan and Chilpansingo; that is, beneath two very distinct porous formations, of which the upper is whitish, argillaceous, and friable, the lower greyish-blue, mixed with great masses of calcareous spar. These two limestones seem at first sight less ancient than the alpine limestone of Peregrino; but they certainly do not belong to the tertiary formations which repose in Hungary on trachytes. I found in them no trace of petrifactions; they are directed N. 35° E., and generally inclined 40°, not at N.W., but at S.E. This uniformity of direction (not inclination,) observed among rocks which appear to be of an age so different, is a very rare phenomenon. It is an additional reason, perhaps, for not considering these porphyries, of which we have just described the position, as trachytes. The limestones of Chilpansingo have cavities that vary from half an inch to eight inches of diameter. The lower formation, which is greenish-blue, immediately covers the porphyry ; it sometimes pierces through the whitish formation, and forms small cylindric or coralliform rocks three or four feet high, at the surface of the soil, which present the most singular aspect. These circumstances of composition and structure indicate a great analogy between the cavernous limestone found from Masatlan and Petaquillas as far as Chilpansingo, and the lower beds of the Jura limestone (höhlenkalk, schlackiger, blasiger kalkstein) which, equally cavernous in the Upper Palatinate (between Lubec and Ettershausen), and in Franconia (between Pegnitz and Muggendorf), give the surface of the soil a peculiar appearance by their asperities. Not far from Zumpango porphyry again occurs below the cavernous limestone of Chilpansingo, or rather below a conglomerate limestone, which, containing large fragments of the blue and white formation, covers the latter in several places. As the porphyries in the groups of Los Caxones and Zumpango rise nearly to the same level (560 and 585 toises), we may suppose, with some probability, that the cavernous limestone which they support in the table-land of Chilpansingo is 800 feet thick.

In advancing to the north towards Sopilote, Mescala, and Tasco, we again lose sight of porphyry. Primitive granite re-appears, but is soon hid by a porphyry the mineralogical composition of which presents very remarkable characters; it is bluish-grey, a little argillaceous by decomposition, and contains large crystals of whitish-yellow feldspar (rather lamellar than glassy), pyroxene of nearly a leek-green, and a little uncrystallised quartz. This stratified porphyry is covered, towards the south, with the same conglomerate limestone that abounds on the table-land of Chilpansingo; towards the north (Sopilote, Estola, Mescala) with a greyish compact limestone traversed by veins of carburetted lime. The limestone of Estola is not always spongy or vesicular in its whole mass, like the formation of Masatlan, but contains large insulated caverns like the limestone of Peregrino, which we have described above. Whilst travelling in those mountains, I had no doubt remaining, that the rocks of Cañada de Sopilote and of Alto del Peregrino are identical with our alpine limestone (zechstein) of Europe, which succeeds, according to the age of its formation, to the red sandstone, or when that is wanting, to the transition rocks. Near Mescala, a little north of Sopilote, rich silver veins, analogous to the veins of Tasco and Tehuilotepec, traverse the alpine limestone.

The rock in the valley of Sopilote, that covers the porphyry of the group of Zumpango, exhibits the same sinuous and contorted beds that are seen at Achsenberg, on the bank of the lake of Lucerne, and in other mountains of alpine limestone in Switzerland. I observed that the upper beds of the formation of Sopilote and Mescala passed progressively to whitish-grey, were destitute of veins of calcareous spar, and presented a dull, compact, or conchoidal fracture. They divide, nearly like the limestone of Pappenheim, into very thin layers. It seems to be the passage from the alpine to the Jura limestone, two formations which immediately cover each other in Switzerland, the Apennines, and several parts of equinoxial America, but which, in the south of Germany, are separated from each other by several interposed formations (by the sandstone of Nebra, or bunte sandstein, by muschelkalk, and the white sandstone or quadersandstein).

Near the village of Sochipala, the alpine limestone is covered by gypsum, and between Estola, and Tepecuacuilco, there appears beneath the alpine limestone (directed sometimes N. 48° E., with inclination 40° at the east; sometimes N. 48° E., with inclination 50° to south-east) a porphyry, asparagus-green, with a base of compact feldspar, divided into very thin strata like that of Achichintla, and almost destitute of disseminated crystals. This rock resembles phonolitic porphyry (porphyr-schiefer) of the trachyte formation. In advancing towards the mines of Tehuilotepec and

Tasco, we find the same rock covered with a quartzose sandstone having a cement of argillaceous limestone, and analogous to the *weiss liegende* (lower arenaceous bed of zechstein) of Thuringe. This quartzose sandstone again announces the proximity of alpine limestone; and in fact, on this sandstone, and perhaps immediately on porphyry (as at Zumpango, and the Alto de los Caxones), near the salt lake of Tuspa, an immense mass of alpine limestone reposes, often cavernous, and containing some petrifactions of trochi, and other univalve shells. This limestone of Tuspa, indubitably posterior to all the porphyries which I have just described, contains beds of specular gypsum, and beds of slaty and carburetted clay, which we must not confound with grauwacke-slate. It is generally bluish-grey, compact and crossed by veins of carbonate of lime. In many places, instead of being cavernous, it passes to a very compact white formation, analogous to the limestone of Pappenheim. I was struck with these variations of texture, which M. de Buch and myself had also observed in the Apennines (between Fosombrono, Furli, and Fuligno), and which seem to prove that where the intermediary members of the series were not developed, the formations of alpine and Jura limestone are more closely connected than is generally admitted. The rich silver veins of Tasco, which formerly yielded 160,000 marks of silver annually, traverse both limestone and a clay-slate that passes to mica-slate; for, notwithstanding the identity of the limestone formations of Tasco and Mescala, which are both argentiferous, the former of those, wherever it has been pierced in mining (Cerro de S. Ignacio), has not been found superposed to porphyry like the limestone of Mescala, but covering a more ancient rock than porphyry, a mica-slate (dir. N. 50° E.; incl. 40°—60°, most frequently at N. W. sometimes at S. E.) destitute of garnets, and passing to primitive clay-slate. It was proper to enter into these details on the formations that succeed porphyries, because it is only in making known the nature of *superposed* rocks, that geognosts can be enabled to decide on the place which Mexican porphyries ought to occupy, in the order of formations. A sketch of a geognostic table has its value only inasmuch as we connect the rock we wish to make known to those by which it is immediately succeeded above and below. Oryctognostic facts alone may be presented singly; positive geognosy is a science occupied with the relation and connection of facts, and in describing any one portion of the globe, we ought not to limit our view, and stop at the study of a particular bed.

Central table-land, valley of Mexico; tract between Pachuca, Moran and La Puebla. An enormous mass of transition porphyry rises to the *mean* height of 1200 to 1400 toises above the level of the sea. It is covered, in the valley of Mexico, and at the south towards Cuernavaca and Guchilaque, with basaltic and cellular mandelstein (*tetzontli* in Mexican); and towards the east and north-east (between Tlascala and Totonilco), with secondary formations. The porphyry, probably hid at first beneath the alpine limestone of Mescala, and then in the Llanos of San Gabriel (near the bridge of Istla) beneath trachytic conglomerates and a porous mandelstein, is identical with that which re-appears, 15 leagues further north, and 800 toises higher, on the banks of the lake of Tezcuco. In the fine valley of Mexico the porphyritic rock pierces the cellular amygdaloid in the hills of Chapoltepec, of Notre Dame de la Guadeloupe, and of Peñol de los Baños. It exhibits several very remarkable varieties: 1°, reddish-grey, a little argillaceous, without distinct stratification, containing crystals of hornblende, and common feldspar in equal parts (gallery dug in the rock of Chapoltepec); 2°, black or darkish-grey (sometimes fissile and spongy) stratified by beds from 3 to 4 inches thick, with a basis of compact feldspar, fracture dull, smooth, or imperfectly conchoidal, resembling more the fracture of lydian stone than that of pitchstone containing small crystals of glassy feldspar and olive-green pyroxene, almost destitute of hornblende, and often covered at their surface by superb masses of reniform hyalite, or Muller's glass (Peñol de los Baños; dir. N. 60° W.; incl. 60° N. E.); 3°, red, earthy, with a quantity of large crystals of common decomposed feldspar (salt-works of the lake of Tezcuco, where the Peñol is covered by ancient Aztic sculptures).

The porphyry of the valley of Mexico furnishes not only springs of pure water, which is conveyed to the town by long and magnificent aqueducts, but but also acidulated thermal waters, some warm, and others cold. Here is also found (and this is a very remarkable fact) naptha and petroleum (promontory of the sanctuary of Guadeloupe), as in the primitive mica-slate of the vicinity of Araya and Cumana. Although this porphyry appears below the porous amygdaloid, and is seen (Cerro de las Cruces and Tiangillo, Cuesta de Varientos and Capulalpan, Cerro Ventoso, and Rio Frio) in all the circular outline of the basin of Tenochtitlan, (the bottom of an ancient lake partly dried,) it is only towards the north-north-east (Pachuca, Real del Monte, and Moran) that it has been found to be argentiferous.

Several rich veins traverse a mass of porphyry above 700 feet in thickness, from the mine of San Pedro, at the summit of Cerro Ventoso (1461 toises) as far as the bottom of the ancient wells of Encino (1170 toises), in the Real de Pachuca. This rock, which would have formerly been called petrosilex, or hornstone-porphyry, is generally greenish-grey, sometimes prase-green, with a scaly fracture, and giving fragments with sharp edges. The paste is probably a compact feldspar, having a large proportion of silica; and contains, not quartz and mica, but crystals of common feldspar and hornblende. In general, the latter substance is not very abundant, and when the porphyry is argillaceous, or merely earthy, we recognise the hornblende only by spots with striated surfaces, and of a very dark green. The beds which are argillaceous and softer (thonporphyry of Moran) appear to be below the harder and more tenacious beds. Strata of phonolite (klingstein) are found subordinate to both; it is smoke-grey, or leek-green, divided into tables or plates that are very sonorous. It is not, however, altogether a porphyry-slate of the trachyte formation, for the phonolitic mass does not contain thin crystals of glassy feldspar, but crystals of greyish-white common feldspar, constantly accompanied with a little hornblende. All these argentiferous porphyries of Moran and Real del Monte are very regularly stratified (general direction as in the valley of Mexico, N. 60° W., incl. 50°—60° at N. E.): they are irregularly columnar only in the Organos of Actopan (Cerro de Mamanchota, summit 1527 toises) and the Monjas of Totonilco el Chico; if indeed the rock of Organos, the mass of which is 3000 feet thick reckoning only the porphyries visible above the neighbouring plains, is identical with the rock of Moran. The latter contains fewer crystals of hornblende; neither of these rocks are fissile or porous, and it is at the foot of the grotesque peaks of Monjas, that the rich veins of Totonilco el Chico are found. The argentiferous porphyries of Pachuca and Moran which I have just described, present no character that should separate them from the transition formation; they are even covered, between the baths of Totonilco el Grande and the cavern of the Madre de Dios, or the pierced rock, by enormous calcareous masses, and by sandstone and gypsum. The calcareous formation, 1000 feet thick, is of a bluish-grey colour, compact, not porous, and contains veins of galena, and beds of white coarse-grained and almost saccharine limestone. It is at least the alpine formation (alpenkalkstein), if it be not a transition limestone; and the relations of position observed between this limestone rock, and the porphyries of Moran and the Magdalena, seem to show that the latter are not trachytic. Advancing four or five leagues distant from the mines of Moran, by Omitlan, by the savannahs of Tinaxas, and a vast forest of oaks towards the Jacal, of which the Oyamel, or *the mountain of Knives* (Cerro de los Navajas), forms the western declivity, we enter a country that exhibits, in its geognostic composition, very recent traces of subterraneous fires. We first find, at the foot of the Oyamel, a greyish-white decomposed porphyry, containing crystals of glassy feldspar, and presenting nearly the same direction (the same angle with the meridian, N. 30° W.) as the argentiferous porphyries, but with an inclination (75° at S. W.) quite contrary. The vegetation prevents fixing the relations of position between the rocks of the Oyamel and the transition porphyries of the silver mines of Moran. The former, which are without obsidian, form the basis of a reddish-white rock, with a lustre like enamel, a smooth fracture sometimes granular, containing a little glassy feldspar, and divided into an infinity of small parallel beds often undulated. This rock is a porphyritic lithoid pearlstone, or rather a trachytic porphyry, not spongy nor fissile, the base of which passes to pearlstone. Such a passage of the base to a mass composed of agglutinated grains is seen even in beds which, from their aspect only, we should at first think were composed of compact feldspar or of a greyish siliceous schist. Neither black mica nor quartz are mixed with the slender crystals of glassy feldspar disseminated in the paste, a mixture which is observed in the pearlstone of Tokai, and of Schemnitz in Hungary.

The abundance of obsidian contained in the porphyries of the mountain of Knives, and through which they are analogous to the pearlstone of Cinapecuaro, leaves no doubt of their volcanic nature. They constitute insulated mountains, sometimes in pairs, with perpendicular beds, resembling by their aspect, hills of basalt, and the trachyte of the Euganean mountains. Did these volcanic masses issue from the transition porphyries of Moran, or does there exist a passage from the one to the other? Are the rocks of Oyamel superposed only on metalliferous porphyries, like the columnar basalts of Regla? We may even enquire if the black, and often cellular porphyries of the valley of Mexico (Peñol de los Baños), covered by amygdaloid which is basaltic and cellular, are of a different origin from the porphyries hid beneath the alpine limestone (Totonilco el Grande)? In this valley of Mexico (advancing from the lake of Tezcuco at the north, towards Queretaro) we see, at Cuesta de Varientos, a decomposed porphyry appearing below volcanic mandelstein, of a brownish red, without hornblende, but abounding in slender crystals of glassy feldspar. On the prolongation of the strata of this rock of trachytic aspect, secondary and tertiary formations repose (Jura limestone, gypsum, and marl with elephants' bones, at 1170 toises high), which fill the basins of Hacienda del Salto, Batas, and Puerto de los Reyes. Ten leagues further, at Lira, we find porphyritic rocks, with a semi-vitreous base, olive-green, covered with mammillated hyalite, and without pyroxene. These rocks contain a little feldspar and grains of quartz, and also small subordinate beds of obsidian. It is, no doubt, a trachyte (a formation in which quartz, even in Hungary, is not entirely wanting). But how shall we distinguish beds of trachyte-porphyry from the transition porphyries by which they are immediately supported, when both, by the mixture of obsidian and pearlstone, have a mineralogical composition so analogous?

This difficulty embarrasses the geognostic traveller still more, when he goes from the valley of Mexico, towards the east, to cross the ridge of mountains on which rise the two volcanoes de la Puebla, Iztaccihuatl (*White Woman*, 2456 toises) and Popocatepetl (*Smoking mountain*, 2770 toises). The porphyritic rocks that appear near the Venta de Cordova and the Rio Frio are closely connected with the trachytes of the great volcano, still burning. They are covered with breccias of pumice and pearlstone with obsidian (between Ojo del Agua and the fort of Perote), and form a basis (between San Francisco Ocotlan, la Puebla de los Angeles, Totomehuacan, Tecali, and Cholula; between Venta de Soto, El Pizarro, and Portachuelo) to a great limestone formation, sometimes compact and greyish-blue, sometimes fine-grained and white, or of a mixed colour. This limestone (transition or alpine) is certainly not tertiary, like the very recent formations of shelly limestone, marl, and gypsum, which, in different parts of the globe, we see placed in insulated portions on the trachytic formations. M. Sonneschmidt saw, near Zimapan, Xaschi, and Xacala, a true transition limestone, darkish-grey and strongly carburetted, resting on porphyries quite similar to those which we have just described in the central table-land of New Spain. Some strata of these porphyries of Zimapan, Xaschi, and Ismiquilpan, contain, like the porphyritic greenstones and pearlstones of Hungary, and like the porphyry placed on clay-slate (of transition?) of the famous mountain of Potosi, garnets disseminated in the mass. They are traversed by veins that contain that magnificent variety of orange-yellow opal, which M. Sonneschmidt and myself made known by the name of fire-opal (feueropal), and which has been found by M. Beudant among the trachytes of Telkebanya. I saw radiated globules of bluish-grey pearlstone, imbedded in the porphyritic paste of Zimapan, and resembling thermantide jaspoid (porcelain-jasper, porzellan-jaspis) in their colour. The relations of position are not yet ascertained, between these porphyries which were considered as trachytic, and those which support the great limestone formations. It is easier to separate metalliferous porphyries from trachytes in artificial classifications, than in the mountains themselves.

Group of porphyries of Guanaxuato. It is this group that determines most clearly the relative age, or, to express myself with more precision, the *maximum* of the antiquity of the Mexican porphyries; if indeed those of which we have just indicated the positions are of the same formation as the porphyries of Guanaxuato. The superposition of those porphyries on rocks belonging to intermediary formation is evident. Near the farm of Noria, and in the Cañada de Queretaro, a slaty olive-green porphyry, containing glassy feldspar in microscopic crystals, is superposed on a transition clay-slate with lydian stone. This superposition is equally certain near Guanaxuato, and particularly near Santa Rosa de la Sierra. The porphyries of this district have in general a concordant position (a parallel direction and inclination) with the strata of clay-slate. They are eminently metalliferous; and the famous vein of Guanaxuato (Veta madre), making the same angle with the meridian as the veins of Zacatecas, Tasco and Moran (N. 50° W.), has been worked successively on a length of 12,000 toises, and a thickness of 20 to 25 toises. It has furnished, in 230 years, more than 180 millions of piasters, and traverses both porphyry and transition-slate. The first of these rocks forms, at the east of Guanaxuato, gigantic masses, that have at a distance the most singular aspect, resembling walls and bastions These perpendicular ridges, rising more than 200 toises above the surrounding plains, bear the name of *buffas ;* they are destitute of metals, appear to have been heaved up by elastic fluids, and are regarded by the Mexican miners, who see them placed also at Zacatecas on a very metalliferous transition clay-slate, as a natural indication of the riches of those countries. When we consider the porphyries of la Buffa de Guanaxuato, and those of the formerly celebrated mines of Belgrado de San Bruno, la Sierra de Santa Rosa, and Villalpando, in the same point of view ; we think we perceive in their latest strata, passages to rocks which, in Europe, are generally placed among trachytes.

In the vicinity of Guanaxuato, porphyries predominate, which have a paste of compact feldspar, are grey and olive-green, and contain imbedded lamellar feldspar (not glassy), either in crystals almost microscopic (Buffa), or in very large crystals (mines of San Bruno and Tesoro). Decomposed hornblende, which probably tinges with green the whole mass of these rocks, is only distinguished by irregular spots. In ascending towards the Sierra (Puerto de Santa Rosa, Puerto de Varientos), the porphyry is often composed of balls with concentric layers; its paste becomes darkish-grey, semi-vitreous (pitchstone-porphyry), and contains a little crystallised mica and grains of quartz. The auriferous veins near Villalpando traverse a green porphyry with a base of phonolite, in which we only perceive some small and thin crystals of glassy feldspar. It is difficult to distinguish this rock from trachytic porphyry-slate; I have seen it covered with an earthy yellowish-white porphyry (mine of Santa Cruz), and with an ancient conglomerate (boca de la mina de Villalpando), which evidently represents the red sandstone, and the lower beds of which pass to grauwacke.

The porphyries of the equinoxial region of Mexico contain, although rarely, besides some disseminated garnets (Izmiquilpan and Xaschi), sulphuret of mercury (San Juan de la Chica; Cerro del Fraile, near the villa of San Felipe; Gasave, at the northern extremity of the valley of Mexico), tin (El Robedal, and la Mesa de los Hernandez), and alumstone (Real del Monte, according to M. Sonneschmidt). This latter substance seems to give to the porphyritic rocks a nearer affinity to real trachytes; although, in South America (peninsula of Araya, Cerro del Distiladero and Chupariparu), I saw a clay-slate which belonged rather to the primitive than to the intermediary formation, traversed by veins, I will not say, of alumstone (alumstein), but of native alum, of which the Indians sell pieces of the size of an inch, at the market of Cumana. The cinnabar of the porphyries of San Juan de la Chica, the argillaceous beds of Durasno containing both coal and cinnabar and placed on a porphyry with much hornblende, are phenomena well worthy of attention. The geognosts who, like myself, attach more importance to the position than to the oryctognostic composition of rocks, will, no doubt, connect the porphyries and clays of Davasno with the deposites of mercury which the formation of the red sandstone and porphyry exhibits in both hemispheres (duchy de Deuxponts and Cuença, between Quito and Loxa). The last beds of the transition formation are found every where in close connection with the most ancient beds of the secondary formation.

The celebrated silver vein of Bolaños contains its greatest riches in an amygdaloid interposed in porphyry. In Hungary, England, Scotland, and even in Germany, rocks of amygdaloid and porphyry belong to grauwacke, to clay-slate, to transition limestone, to red sandstone, and to coal sandstone. The metalliferous porphyry of Guanaxuato simply covers clay-slate, without forming in it subordinate beds (as in the group § 22); but a syenite analogous to that which occurs in the mine of Valenciana, in the midst of intermediary clay-slate, alternates thousands of times, on a surface of more than twenty square leagues, with transition greenstone, between the mine of the Esperanza, and the village of Comangillas. In that region the syenitic rock is destitute of metals, but at Comanja it is argentiferous, as it is also in Saxony and in Hungary.

b. *In the austral hemisphere.* Between the 5th and 8th degrees of latitude I saw porphyritic rocks, intimately connected together, covering the eastern and western declivities of the Andes of Peru. These rocks repose either on mica-slate (of transition ?) traversed by argentiferous veins (Mandor, El Pareton), or when mica-slate is wanting, upon granite. Some are divided into immense columns (Paramo de Chulucanas), or very regularly stratified (Sondorillo). Their base is black and almost basaltic; they contain more pyroxene than feldspar, and alternate (Quebrada de Tacorpo) with beds of jasper and compact feldspar. The latter, destitute of disseminated crystals, is black, like lydian stone, and reminds us by its colour and homogeneous nature, of some of the basanites used by the ancients. Other porphyries (N. S. del Carmen, at the north of the Indian village of San Felipe) have a less trachytic appearance, furnish rich silver veins, and are covered sometimes by beds of quartz three or four toises broad, sometimes by compact limestone (alpine?), darkish-blue, traversed by small veins of calcareous spar, and filled with petrified shells (hysterolites, anomiæ, cardia, and fragments of large polythalamous shells, which are rather nautilites than ammonites). In descending (still on the eastern declivity of the Andes) towards Tomependa, on the banks of the river Amazon, I saw, between Sonanga and Chamaya, the old red sandstone (todtes liegende) superposed on an earthy greyish porphyry, containing (like that of Pucara) a great quantity of hornblende, and a little common feldspar. We find, on the western declivity of the Andes, in approaching the coast of the South Sea (between Namas and Magdalena) porphyries altogether destitute of hornblende, and supporting that great formation of quartz which replaces the red sandstone in this region. I stated above (§ 18.), that this porphyry, far from being primitive, appeared to me to be the most ancient of the transition porphyries. This result can only be stated doubtfully; for, between Ayavaca, Zaulaca, Yamoca (§ 8.), and Namas (province of Jaen de Bracamoros, and the intendance of Truxillo), it is very difficult to determine with certainty the age of the granite, syenite, and clay-slate, on which the intermediary porphyries or porphyritic trachytes repose. When the relations of superposition are not exactly known, we ought to be cautious in pronouncing on a formation, the geognostic constitution of which is so complicated.

B. *Group of Hungary.*

The formation of syenite and porphyritic greenstone contains the principal mineral riches of Hungary and Transylvania (Schemnitz, Kremnitz, Hochwiesen, and Kœnigsberg; the Bannat, Kapnak, and Nagyag). We shall describe this formation from the observations of M. Beudant, which are not yet published.* The formation of Hungary is

* M. Beudant's work has since been published. Trans.

much less simple than that of Mexico, with which, however, it is found to have a great analogy. The rocks that constitute its principal mass are porphyritic rocks with a base of compact feldspar coloured green; they contain hornblende like the porphyries of equinoxial America, which I have described above, and are almost destitute of quartz. This latter substance appears only in the subordinate beds of syenite, granite, gneiss, and compact greenstone, to which the porphyritic rock passes. In New Spain, the porphyries with gold and silver veins have a paste apparently homogeneous, most frequently slightly coloured; in Hungary, porphyric greenstones, and not true porphyries, are the predominating rocks. According to simple oryctognostic considerations, that is to say, those of composition, the auriferous formation of Hungary resembles much more the Mexican formation of Ovexeras, in which syenites and greenstone, more or less porphyritic, alternate, than those great masses of porphyry that traverse the celebrated veins of Pachuca, Real del Monte, Moran and Guanaxuato (at the south-east of the mine of Belgrado); but, geognostically considered, all those porphyritic and syenitic rocks of Mexico and Hungary constitute but one formation, sometimes simple, sometimes composed (with alternations).

The porphyritic and syenitic rocks of Hungary, the most compact as well as the most mixed, contain carbonate of lime, and effervesce with acids. This character is found in the Mexican rocks in an analogous position, but not in the superimposed trachytes. Glassy feldspar is much more rare in the porphyries with a greenstone base of Hungary, than in the Mexican porphyries; it occurs only (Hochwiesen, Bleihütte) in the upper and earthy strata, particularly where the trachytic formation commences. Oxidulated iron abounds where the hornblende appears in very distinct crystals; garnets (which we have already mentioned above in the Mexican porphyries of Zimapan, and those of Potosi, on the eastern ridge of the Andes of Peru) penetrate to the middle of the prisms of hornblende. Although, in the great formation of syenite and porphyritic greenstone of Hungary, the varieties of rock pass frequently into each other, we may observe, in general, the following type of association and of superposition; the lower part of the whole system is formed by syenites, coarse and fine grained, passing to a talcose granite (Hodritz) and to gneiss; the middle part is composed sometimes of compact greenstone, with a black paste almost destitute of disseminated crystals, sometimes of porphyritic rocks with a base of pure feldspar, or a base of feldspar and hornblende mixed, containing imbedded crystals of common feldspar (lamellar), hornblende, a little mica and garnets, and very rarely quartz; the upper part exhibits porphyritic greenstone decomposed and extremely auriferous. It is only this last layer that sometimes contains glassy feldspar, laumonite, mica, and (as in equinoxial America) veins of red jasper. In the earthy greenstones, which are of a more simple structure because they do not alternate with syenite, granite, or transition gneiss; compact basaltiform masses are found (valley of Glashütte) divided into prisms, and a black porphyritic greenstone with a base of hornblende and feldspar. This greenstone contains very small acicular crystals of hornblende, numerous *laminæ of black mica*, and *druses* of white and red quartz.

The beds subordinate to the great formation of syenite and porphyritic greenstone of Hungary, are mica-slate (valley of Eisenbach); compact quartz, sometimes laminar and micaceous, sometimes granular passing partially to a dull flint with a smooth fracture (western basin of Schemnitz); steatitic limestone, sulphur-yellow, greenish, or reddish, with garnets disseminated in the mass, and accompanied by serpentine (Hodritz). All this system of syenitic and porphyritic rocks is very distinctly stratified in Hungary as well as in Mexico; but in the former of those countries the direction and inclination of the strata are uniform only in the same group of mountains. It is not easy to determine with certainty the nature of the formation on which the syenites and porphyritic greenstones of Hungary repose. M. Beudant believes that they are of a more recent formation than grauwacke, which is not developed in Hungary where the porphyritic greenstone predominates. Talc-slate, alternating with greyish crystalline limestone, and probably belonging to the most ancient transition formation, appeared to that learned geognost, as well as to M. Becker, to serve as the basis of the syenitic and porphyritic formation. This analogy may be added to the others which this formation presents with the homonymous formation of Mexico. In Hungary, as well as in the New Continent, porphyry, syenite, and greenstone are immediately covered by trachytes and trachytic conglomerates containing obsidian and pearlstone. In Auvergne (Mont-d'Or, Cantal); in the Greek islands (Argentiera, Milo, Santorino), visited by an excellent observer, M. Hawkins; at Unalaska, recently explored by M. de Chamisso, and by the expedition of Captain Kotzebue, the same relations of position are observed between trachytes and transition porphyries. At the mountain of Kasbeck, in the Caucasian chain, an intermediary porphyry, which alternates with syenite, granite, gneiss, and transition clay-slate, contains also glassy feldspar; it even presents in some strata every appearance of a porous trachyte. Thus, on the most distant points of the globe, in America, Europe, and Asia, we see that porphyries oscillate between transition rocks and very ancient volcanic rocks.

C. *Group of Saxony.*

We do not here speak of the porphyry which forms with greenstone and darkish-grey limestone (Friedrichswalde, Seidwitzgrund) subordinate beds in transition slate (§. 22.), but of the great formation of syenite and porphyry which Werner designated by the name of the *principal formation*, (*Hauptniederlage*). That illustrious geognost distinguished four formations of porphyry; the first forming beds (or rather veins?) in gneiss and primitive mica-slate; the second alternating with syenite; the third belonging to the coal sandstone, and containing greenstone, pitchstone, and agathiferous amygdaloids; the fourth, interposed in trap rocks (volcanic). These four formations, the first of which does not probably constitute an *independent* formation, are, as I have elsewhere stated (*Voyage to the Equinoxial Regions*, t. i. p. 155.), porphyries interposed in primitive rocks, transition porphyries, secondary porphyries, and trachytes (trapporphyre). The *principal formation* of porphyry and syenite of Saxony reposes on transition slate (with grauwacke), and consequently, where clay-slate is not developed, on more ancient rocks. The syenite, which alternates with porphyry (Meissen, Leuben, and Prasitz; Suhl) passes into granite and gneiss. Transition granite is generally coarse-grained, composed of reddish feldspar, smoke-grey quartz, and black mica well crystallised (Dohna, Posewitz, and Wesenstein). Transition gneiss (Meissen) is more rare than granite, and forms beds in syenite, like granular and white limestone (Naundorf) and greenstone which passes to basalt (Wehnitz). The presence of the syenite formation, which contains, in the valley of Plauen (as in Norway), some disseminated crystals of zircon, is often manifested only by beds of granite; for, the frequent and local substitution of mica for hornblende, and of hornblende for mica, characterises the syenitic formation abounding in brown sphene *(braunmen-akanerz)*, which is a silicate of titanium and lime. The unstratified porphyry of Saxony has generally a base that is red, greyish, argillaceous (thonpor-phyr), the result of the decomposition of the compact feldspar; this base, according to M. Boué, sometimes takes the aspect of clinkstone (klingstein). This porphyry contains scarcely any hornblende, and is not without quartz, like that of Mexico and Hungary. We find in it common feldspar, quartz crystallised in double hexahedral pyramids, and sometimes a little mica. The group of porphyries and syenites of Saxony are somewhat metalliferous; the syenite stratified in thick beds of Scharfenberg affords veins of silver, and the porphyry of Altenburg sometimes contains tin.

In the valley of Plauen, near Dresden, is found the rock to which Werner first gave the name of syenite, believing erroneously that all the Egyptian obelisks preserved at Rome contained hornblende, M. Wad (*Foss. Ægypt. Musei Borgiani*, 1794, p. 6. and 48.; Zoega, *de Obeliscis*, p. 648.) has proved that these obelisks, the finest of which, speaking mineralogically, is that of the Piazza Navona, are of true granite, with black agglomerated mica, without hornblende. In fact, there does not exist any formation at Syene independent of syenite and intermediary porphyry; but primitive granite, not perhaps of a very ancient formation, contains there hornblende (as at the Oroonoko, at Spitzberg, near Krummhübel in Silesia, and near Wiborg in Finland) disseminated in subordinate beds of little extent, and extending irregularly. In a classification of rocks, the rock of Syene should be considered as a granite containing hornblende, and not as a syenite. Some insulated fragments of that rock, which are found among Egyptian monuments, deceived Werner by the oryctognostic analogy which they present with the syenite of the valley of Plauen.

Formations of porphyry and syenite, entirely similar to that of Saxony, and placed on transition slate and grauwacke, are common at Thüringerwald in Moravia (between Blansko, Brünn, and Znaim), according to M. Boué; and according to M. Rozière, in the peninsula of Mount Sinai. The latter merit particular attention. Intermediary schistose and arenaceous rocks cover a part of Arabia Petrea. In the midst of those rocks which contain conglomerates with fragments of granite and porphyry (*universal breccia of Egypt*, in the language of antiquaries), syenites occur, and porphyries with a base of siliciferous compact feldspar, containing imbedded crystals of lamellar feldspar, a little hornblende, and also quartz, according to M. Burckhardt. The porphyries are generally inferior to the syenite, and the latter, of which the *tables of the law* were probably composed, and which it is believed are buried at Djebel Mousa, is accompanied with blackish compact greenstone (gulph of Akaba) and porphyritic greenstone. All this formation in Arabia Petrea, of which I have examined numerous specimens, resembles in the most striking manner, the porphyritic and syenitic formations of Ovexeras and Guanaxuato, in Mexico. By substituting with M. Rozière, the word *sinaite* for that of syenite, we should have given to the transition rocks, which are composed of hornblende and feldspar sometimes mixed with a little quartz and mica, a more exact *geographical name* — a name which (like that of Jura limestone) not only recals the relations of composition, but also those of position.

D. *Group of Norway.*

§ 24. This is the series of rocks described by two celebrated geognosts, professor Haussmann, and M. Leopold de Buch; it is this in which the formation of granite posterior to calcareous rocks containing the remains of organised bodies is best developed, and which has consequently thrown much light on the true nature of the rocks of transition. This class of rocks was at first considered only as an association of grauwacke, carburetted schist, and black limestone; but by degrees it was observed, that the great mass of porphyries, long called primitive porphyries, belonged either to the transition formation, or even to the red sandstone. The syenites of Meissen had been united to the intermediary porphyries; but although the former want hornblende, and pass insensibly to transition granite (Dohna), this phenomenon, the new appearance of granitoid rocks altogether analogous to primitive rocks and covering both black porphyries with pyroxene and orthoceratite limestone, began to engage the attention of geognosts only when the shores of the gulph of Christiania were described in their wonderful relations of superposition.

The zircons, which have given so much celebrity to the syenite of Holmstrand, and of Stromsoë, are found abundantly in the syenites of South Greenland (according to M. Giesecke, near Cape Comfort, at Kittiksut and Holsteensberg); they are also disseminated, in very small masses, in the syenites of Meissen, and of the valley of Plauen. This substance, in other localities, belongs rather to primitive rocks (for instance, to gneiss); for, although zircon, titaneous iron, sphene, epidote, glassy feldspar, chiastolite, lydian stone, diallage, hornblende, and pyroxene, particularly abound in certain formations, we must not consider those associations as characters of absolute value. The accumulations of zircons in the syenites of Christianiafiord is much less remarkable with respect to geogonic problems, than the number of vacuities, the cavernous and fissured structure of those very transition syenites that are connected with basaltic and pyroxenic porphyries. Since, by the frequent analogies observed between the porphyry and syenite formation of Christiania, and the transition formations of Caucasus, Hungary, Germany, the west of France, Greenland, and Mexico, geognosts are no longer astonished at feldspathic and crystallised rocks succeeding to grauwacke and limestone with entrochi and orthoceratites, the appearance of those very crystallised rocks, in the most ancient member of the series of secondary rocks, begins to engage their attention. Crystalline masses have been observed in both hemispheres, composed of feldspar and hornblende or feldspar and pyroxene, *oscillating* between volcanic formations, the intermediary formation, and the red sandstone. These *oscillations*, these intercalations of problematic rocks, which we are tempted to regard as the effects of a successive penetration from below upwards, prove the close connection that exists between the most recent transition beds and the most ancient secondary and volcanic rocks. In the southern part of the Tyrol, masses of granite and syenitic porphyry appear to break out from the red sandstone, in the alpine limestone; and these curious phenomena of *alternation*, connected with so many others previously known, appear to forbid both at the separation of the coal sandstone from the porphyries of the intermediary formation, and the *historical*, and too exclusive *denomination* of pyrogenic rocks.

The great formation of the porphyries, syenites, and granites of Norway, reposes on a formation of transition slate, containing alternately, beds of black limestone, lydian stone, and perhaps even (for the position in this part is less evident) of granite. The black limestone (Aggerselv, Saasen) contains orthoceratites, several feet long, with entrochi, madrepores, pectinites, and (although very rarely) ammonites. Veins of porphyry and porphyritic greenstone, from two to fifteen toises thick, traverse clay-slate and limestone (Skiallebjerg), and announce the approach to analogous masses of porphyry that repose, not immediately on clay-slate, but on an arenaceous rock (grauwacke) by which the clay-slate is covered. Between Stromsoë, Maridal, and Krogskovn, grauwacke, instead of being found in beds in clay-slate, to which it belongs (§ 22.), forms a kind of upper layer, so that it is there seen to change from below to above; primitive gneiss, transition clay-slate, alternating with orthoceratite limestone; grauwacke, porphyry, with subordinate beds of greenstone; granite and syenite, with zircons, alternating with some beds of porphyry. Near Skeen and Holmstrand, the orthoceratite-limestone acquires so great a development, that clay-slate is there entirely wanting; and grauwacke is replaced by a bed of micaceous quartz. We there see from below upwards, primitive gneiss, transition limestone, quartz-rock; porphyry of which the lower bed is of mandelstein; and syenite with zircons. The porphyries of Christianiafiord, mixed by infiltration with carbonate of lime, are generally reddish-brown; they have sometimes very slender crystals of lamellar feldspar, and are almost destitute of quartz and hornblende. Crystallised quartz occurs only between Angersklif and Revo. The paste of the porphyry becomes occasionally black and porous (Viig, Holmstrand). In this state, the rock resembles basalt, like the syenite of the peninsula of Mount Sinai, and contains crystals of pyroxene. M. de Buch, to whom I owe all these important facts, observes, that crystals of feldspar disappear in proportion as the mass takes a darker tint, a phenomenon which I also remarked in several transition porphyries of Mexico. Mandelstein, of which the elongated cavities are filled with carbonate of lime, and which forms the inferior bed of the Norwegian porphyries of Skeen and Klaveness, reminds us of the mandelstein of the porphyry of Bolaños (Mexican province of New Galicia), which is traversed by one of the richest argentiferous veins. The *syenites* of Christianiafiord, always placed above porphyries, although alternating with them at first, are composed (Waringskullen, Hackedalen) of a great quantity of large crystals of red feldspar, and a little hornblende in very small crystals; mica and quartz are only accidental. Some angular cavities in the syenite contain crystals of zircon and epidote. The ferriferous titanium, common in both hemispheres to primitive rocks of euphotide and trachytes, are found occasionally in the mass of syenites with zircons.

VI. TRANSITION EUPHOTIDE.

§ 25. We must here distinguish, as among the syenites, between interposed beds and independent formations. Beds of serpentine are found interposed in whitestone (§ 4.), in primitive mica-slate (§ 11.), and in transition clay-slate (§ 22.). With respect to independent formations of euphotide (gabbro), which are often of a very complicated structure, we may at least reckon two, even if we reject the formation not covered by any other and rather doubtful, of Zöblitz in Saxony. The first of these independent formations occurs (§ 19.) on the limit of primitive and intermediary formations. It is that which M. de Buch has made known in Norway (Maggeroe, Alten), and M. Beudant in Hungary (Dobschau). The second formation belongs to the newest transition rocks, and occurs on the limit of the intermediary and secondary rocks. The serpentine connected with the formation of *ophite*, observed by M. Palassou in the Pyrenees (valley of Baigorry, Riemont), in the department of Landes, has been considered as more recent. But this ophite is a greenstone, an intimate mixture of feldspar, epidote, and hornblende, in which beds of serpentine are interposed (Pousac); it passes by the change in the proportion of the elements, sometimes to syenite, sometimes to graphic granite. M. Boué, who has recently examined this ophite on the spot, believes it to be a transition formation, covered by variegated sandstone, clay, and secondary gypsum.

The great formation of transition euphotide (that which constitutes the last member of the series of intermediary rocks) seems almost constantly connected (as in Piedmont, between Mont Cervin and Breuil) with hornblende rocks. On the northern limit of the Llanos of Venezuela, we see, between Villa de Cura and Malpasso, considerable masses of serpentine, covered by red sandstone, reposing on green clay-slate and transition limestone, and sometimes immediately on primitive gneiss. A small-grained greenstone forms beds both in clay-slate and serpentine. The latter is even sometimes mixed with feldspar and hornblende. Green and blue schists, greenstone, black limestone, and serpentine traversed by veins of copper, form but a single formation, which is covered and intimately connected with pyroxenic amygdaloids, and phonolite. I have described the remarkable position of the serpentine rocks of Venezuela, in the 16th chapter of my *Voyage to the Equinoxial Regions of America.*

In the island of Cuba, the bay of Havannah separates Jura limestone from a euphotide formation, the lower beds of which alternate, not with greenstone, but with a real transition syenite, composed of a great quantity of white feldspar, decomposed hornblende, and a little quartz. The alternating strata of syenite and serpentine are sometimes of the thickness of three toises; the upper layer of this mixed formation is serpentine, forming hills thirty to forty toises high, abounding in diallage metalloïd, and traversed by veins of fine calcedony, amethyst, and ores of copper. This rock is confusedly stratified (by groups, N. 55° E.; incl. 60° at S.W. or N. 90° E.; incl. 50° at N.); springs of petroleum and water impregnated with sulphuretted hydrogen issue from thence.

To this formation of transition euphotide (§ 25.), the formation of Scotland (Girvan and Bellantrae), composed, according to M. Boué, of serpentine, hypersthene rocks, and syenite, seems to belong, as also the celebrated formation of Florentine, (Prato, Monteferrato), described by MM. Viviani, Bardi, Brocchi, and Brongniart. Hypersthene often replaces diallage (Scotland, and Gernerode in Germany). With respect to the euphotides of Florence, they have been lately the subject of some interesting discussions. They contain beds of reddish jasper, sometimes striped, and which appear superposed, according to M. Brocchi, like those of Styria, on grauwackes and transition limestone. M. Brongniart thinks that the arenaceous formation, or, as he calls it, the calcareo-psammitic formation of the Apennines, which serves as the basis of jaspiferous euphotide, is either a very ancient secondary rock, or a very modern transition rock. This geognost has made known the near connection that exists between the serpentine of Italy and the jasper formation. The latter generally constitutes the lower strata of the euphotides.

Here the series of intermediary formations terminates. We have enlarged the more in their description, because in attempting to exhibit them, according to a new mode of classification, by groups, we wished to fix the attention of geognosts on various phenomena of position which occur in the mountains of Mexico and South America, which are as yet but little known.

SECONDARY FORMATIONS.

I. *The great deposite of coal, red sandstone, and secondary porphyry* (amygdaloid, greenstone, pitchstone).

II. *Zechstein* (alpine limestone, magnesian limestone), sometimes interposed in the red sandstone, hydrated gypsum, rock-salt.

III. *Alternating deposites, arenaceous and calcareous* (marl and oolite), placed *between the zechstein and the chalk.* We shall here only mention two types very similar in their geognostic relations, beginning each series by the most ancient rocks.

FIRST TYPE.

Grès bigarré, variegated sandstone (with oolite) and *clay* with fibrous gypsum and traces of rock-salt.

Muschelkalk (limestone of Gottingen). *Quadersandstein.*

Jura limestone, in several beds of porous and cellular limestone; marly limestone, with bones of the ichthyosaurus (lias); oolites, limestone with madrepores (coral rag), limestone with fossil fish and crabs.

Clay with lignite.

Green sandstone and sand (chalk with chlorite, or plänerkalk).

SECOND TYPE.

Red marl, marl formations with gypsum and rock-salt.

Oolite formations, of which the lower bed is the lias. Green sand, which represents chalk with chlorite (craie chlorite).

IV. *White and grey chalk, or craie tuffeau.*

TERTIARY FORMATIONS.

Deposites above the chalk. Their order of succession differs according to the alternation of the partial formations which are found more or less developed. We present the most complicated and the best known type: *Plastic clay with lignite,* amber, and quartzose sandstone. (A formation nearly parallel, and perhaps still newer, is that of the molasse and nagelfluhe of Argovia, with lignite and fossil bones.)

Coarse *limestone* (calcaire grossier) *of Paris.* The upper and beds are of sandstone.

Marl and gypsum with bones. The lower strata are of siliceous limestone.

Sandstone and sands of Fontainbleau, upper fresh-water formation. (Siliceous millstone; limestone of d'Œningen, perhaps connected with the molasse. Travertino.) Alluvial deposites.

FORMATIONS EXCLUSIVELY VOLCANIC.

I. *Trachytic Formations.*

Granitoid, and syenitic trachytes.

Porphyritic trachytes (feldspathic and pyroxenic).

Phonolites of the trachytes.

Semi-vitreous trachytes.

Pearlstone with obsidian.

Trachytic meuliere, cellular, with siliceous nodules.

(*Trachyte and pumice conglomerates,* with alum-stone, sulphur, opal, and opalised wood.)

II. *Basaltic Formations.*

Basalts, with olivine, pyroxene, and a little hornblende.

Phonolites of basalts.

Dolerites.

Cellular mandelstein.

Clay with pyrope-garnets. This latter formation seems connected with the clay with lignite of the tertiary formation, upon which the basaltic currents have spread.

(*Conglomerates and basaltic scoriæ.*)

III. Lavas that have issued from a volcanic crater. Large currents of ancient lavas, generally abundant in feldspar. Modern lavas in distinct currents, and not broad; obsidian, and obsidian-pumice.

IV. *Volcanic tufas,* with shells.

(Deposites of compact limestone, marl, gypsum, and oolites superposed on the most modern volcanic tufas. These small local formations belong perhaps to the tertiary formations. Table-land of Riobamba, Fortaventura and Lancerote islands.)

I have stated above, the reasons which led me to make secondary and volcanic formations succeed at once, as by bisection, to the transition formations. The latter are connected by their grauwackes and porphyries, as well as by a great accumulation of carbon, to the red sandstone, secondary porphyries, and deposites of coal; and they are connected by their porphyries and syenites to the trachytes. These connections are so strong, that it is often difficult to separate porphyries, amygdaloids, and pyroxenic rocks belonging to the transition formation, either from the red sandstone with interposed beds of porphyry and greenstone, or from those formations which are exclusively volcanic. I use the expression *exclusively volcanic formations*, to remind the reader, that beyond this formation there may be igneous rocks, but that we no where else find a less interrupted, and less contested succession.

SECONDARY FORMATIONS.

These formations are very unequally developed on the globe, and the cause of this inequality of developement is one of the most interesting problems of *geogony* or *historical geology*. We rarely find all the members of the series of secondary and tertiary formations together in the same country (Thuringia, Hanover, Westphalia, Bavaria, south of France, centre and south of England); considerable formations, for instance, red sandstone, or alpine limestone, are often entirely wanting; sometimes the second is contained in the first like a subordinate bed; at other times, all the terms of the geognostic series, between the alpine and Jura limestone, or those which are posterior to the chalk, are suppressed. In the Scandinavian peninsula, on the coast of the sea of Behring, and (if we except the sandstone with lignites which covers the basalts), even in Greenland, this suppression extends over all the secondary and tertiary formations. It was long believed that this singular phenomenon was confined to the frigid zone, and particularly to that part contained between the 60° and 70° of latitude; but I have also found in a vast space of the Sierra Parima, near the equator, between the basin of the Amazon, and that of the Lower Oroonoko, (lat. 2°—8°, long. 65°—70°) the primitive formation of granite-gneiss not covered by intermediary, secondary, or tertiary formations. Where there is a complete deficiency of the formations posterior to the developement of organised beings on the globe, limestone oftener than sandstone formations are wanting, for every formation not schistose has breccias or conglomerates which are peculiar to it. These conglomerates are small partial deposites that must not be confounded with the great independent formations of grauwacke, old red sandstone, variegated sandstone, and quader-sandstein.

I. COAL, RED SANDSTONE, AND SECONDARY PORPHYRY, WITH INTERPOSED AMYGDALOID, GREENSTONE, AND LIMESTONE.

§ 26. The coal sandstone and porphyry constitute one formation (rothes todtes liegende), of a variable aspect, and often of a very complicated structure. Cellular mandelstein, greenstone, feldspathic and pyroxenic granular rocks, retinites (pechstein), and some fetid limestones belong to this formation as interposed beds. The English mineralogists call a formation of red sandstone and porphyry by the name of *new red conglomerate* (Exeter and Teignmouth), to distinguish it from their old red sandstone of Mitchel Dean in Herefordshire, which is an arenaceous transition rock, (grauwacke), placed between two transition limestones, those of Derbyshire and Longhope. This nomenclature, which the learned professor of Oxford, Mr. Buckland, has lately explained, has occasioned many geological mistakes. I believe it would be very useful to the progress of the science of positions, if by degrees, those vague denominations were abandoned of *ancient, intermediary, and new sandstone, lower* and *upper* sandstone, and gypsum, and limestone of the *first, second,* and *third* formation. They are only relatively true, in particular places; and they enumerate what is numerically variable, according to the alternations and suppressions of different terms of the series.

The transition formation not only affords anthracite, but also true coal. We find small deposites in England in the old red sandstone (Bristol), the lower of which beds pass from a fine and marly conglomerate to a very compact grauwacke; and in the mountain limestone (Cumberland), which is analogous to the transition limestone of Namur in Belgium, and Prague in Bohemia. But the great deposite of coal (coal measures) occurs, as we have said above, on the limit of intermediary and secondary rocks. On account of this very position, the coal is sometimes mixed (England, Hungary, Austria, south of the Danube, Belgium) with arenaceous beds connected with true grauwackes; sometimes (and it is the type most generally known on the Continent, since the observations of Fuchs and Lehman, made towards the year 1750) it belongs to the great formation of porphyry and red sandstone. In the first case (England), the deposites of coal follow the inclination of the transition rocks, to which (as MM. Conybeare and Phillips have judiciously observed), they are more particularly connected and are found as much inclined as the black lime-stone and grauwackes which they cover. The series of horizontal and secondary formations there appear to begin only with magnesian limestone, which represents the zechstein or alpine lime-stone. In the second case (Germany and France), the coal deposite accompanies the red sandstone and porphyry, whatever may be the primitive or intermediary formations on which those two rocks are immediately placed. This constant union with superposed rocks, and want of connection with the lower formations, are the surest geognos-tic characters of the dependence or independence of a formation. The great deposite of coal is often neither covered with porphyry, nor red sandstone, nor mixed with arenaceous beds belonging to in-termediary formations. It is often placed in ba-sins surrounded by hills of sandstone and porphyry, and contains at its upper part only alternating beds of slate-clay (schieferthon), sometimes blu-ish-grey, friable and filled with impressions of fern, sometimes compact, carburetted (brand-schie-fer), and pyritous. These strata of carboniferous sandstone (kohlen-schiefer), quartzose sandstone passing to granular quartz, conglomerates in large fragments (steinkohlen-conglomerat), and fetid limestone, are found amidst slate-clay before reach-ing the coal. These are small local formations contained in the deposites of muriatiferous clay (salzthon), rock-salt, hydrate of iron, and calamine; and in circumstances quite analogous, where they are not immediately covered by the great formation of alpine limestone. Notwithstanding these appearances of being insulated and independent, coal and rock-salt do not the less belong, geognostically, one to the red sandstone, and the other to alpine limestone or zechstein. The impressions of arborescent ferns, as MM. Voigt and Brongniart have well observed, characterise the epoch of true coal, while they do not appear in the lignites.

In the temperate zone of the old continent coal decends as far as the lowest parts of the shore. We find at Newcastle on Tyne, at the level, and below the bottom of the sea, fifty-seven beds of hardened clay and conglomerate, alternating with twenty-five beds of coal. In the equinoxial region of the new continent, I saw, on the contrary, coal interposed in the red sandstone, and rising in the table-land of Santa Fé de Bogota (Chipo between Canoas and Salto de Tequendama; mountain of Suba; Cerro de los Tunjos), to 1360 toises above the level of the ocean. Coal is also found in the southern hemisphere, in the lofty Cordilleras of Huarocheri and Canta; I was well assured that they occur near Huanuco (interposed with alpine limestone,) very near the limit of perpetual snow, at the height of 2300 toises, consequently above all all phanerogamous vegetation. Deposites of coal abound beyond the tropics in New Mexico, in the middle of the salt plains of Moqui and Nabajoa, and to the east of the rocky mountains, as also towards the sources of the Rio Sabina, in that immense basin covered with secondary formations, in which flow the Missouri and the Arkansas. Rhomboidal fibrous masses, having a silky lustre, and staining the fingers, are found imbedded in the compact coal of both continents; they form a kind of breccia, which the miners consider as containing fragments of carbonised wood; sometimes the shining masses are almost incombustible, and seem a kind of anthracite with a fibrous texture (faserkohle d'Estner, mineralische holzkohle of Werner). They are found, according to the observations of MM. de Buch and Karsten, collected (Lagiewnick in upper Silesia) in beds from four to five inches thick. This phenomenon merits particular attention; for the coals which contain the imbedded fragments with a silky lustre, belong to the best-characterised red sandstone, and not to the lignites of the clay immediately below or above the chalk. In the peninsula of the Crimea a vast district presents numerous alternations of beds of slate-clay without coal, conglomerates, greenstone, and compact limestone. Is this a formation of red sandstone, containing hornblende rocks, and alternating with the zechstein?

It is difficult to assign a general type to the different divisions that constitute the great formation § 26. The coal appears most frequently below the red sandstone; sometimes it is placed evidently in that rock, or in porphyry. The porphyry penetrates and abounds in different manners in the formation of coal sandstone; sometimes covering immediately the coal; more generally surmounting the sandstone, rising in domes, or rocks with steep declivities. When transition formations are immediately covered by the red sandstone (Saxony) it is often difficult to decide, if the porphyries found in the proximity of coal are transition porphyries, or if they belong to the red sandstone. It appears also that porphyries form real beds less frequently than crossing and intermingled masses (stehende Stöcke and Stockwerke) in the coal formation. They vary much in colour; they are violet, grey, and reddish-brown, verging on white (Petersberg near Halle, Giebichenstein, Wettin), penetrated by fluate of lime unstratified, sometimes divided into thin plates, and accompanied by *porphyritic breccias.* The base, or paste of those porphyries, besides containing imbedded lamellar feldspar that is sometimes steatitic, blackish quartz, and a little brown mica and hornblende, is generally formed of compact feldspar. This paste passes to kaolin (Morl near Halle); at other times it becomes black and almost basaltic (Lobegün in Saxony, Schulzberg in Silesia), cellular, and as if scorified (Plizgrund near Schmiedsdorf in Silesia), or passing to phonolite (Zittau in Saxony). The same analogies are sometimes remarked in the porphyries, amygdaloids, greenstones, and pyroxenic rocks of the red sandstone (Saxony, Silesia, Palatinate, Scotland), with the rocks exclusively called volcanic, as are found in the porphyries and syenites of intermediary formation (Hungary, Norway, Mexico, Peru). M. de Buch saw in Silesia porphyries of the red sandstone abounding in crystals of hornblende (Reichmacher near Friedland), or containing also imbedded quartz (Wildenberg near Jauer), and slender crystals of glassy feldspar. M. Boué observes, that in the red sandstone of Scotland, which, in general, is almost destitute of coal (with the exception of Dumfries), the interposed trap-rocks have elongated cavities with a shining surface. This porous mandelstein of the red sandstone has all the appearance of intercalated volcanic rocks.

Germany, at its northern extremity (isle of Rugen), contains chalk and tertiary formations, and at its southern, in the Tyrol (valley of Eisack, Collmann, Botzen, Pergine, Neumarkt), porphyries of the red sandstone. The composition of these porphyries of the Tyrol is identical with that of the porphyries of Mansfeld; they contain, besides feldspar, black mica, brown quartz, and a little hornblende. The red colour of their paste sometimes penetrates within the imbedded crystals of feldspar. In a geognostic journey made in 1795, I found these porphyries somewhat regularly stratified near Botzen and Brandsol (N. 25° W.; incl. 30° at S.E.). They present small deposites of coal on the branches of the Adige, between Saiss and S. Peter.

Secondary porphyries exhibit, in every part of Europe, the appearance of a progressive passage to the red sandstone. Some geognosts admit that insulated crystals of feldspar occur imbedded in the cement of the arenaceous rock, or are developed there; others assert, perhaps with more reason, that these pretended passages of porphyries to porphyritic breccias and red sandstone, are only the effect of an illusion produced by *regenerated porphyries*, that is, by agglomerates which are formed at an epocha when the imbedded fragments were in a state too soft to preserve their outlines amidst the cement. A porphyritic breccia (trümmer porphyr) near Duchs in Bohemia, which M. Freiesleben and I described in 1792, and in which grains of quartz are mixed with broken crystals of quartz and feldspar, may throw some light on a phenomenon which has not yet been sufficiently investigated. It is very remarkable, and this observation has been long since made, that porphyries are wanting on the north of the Alps of Switzerland and the Tyrol, while they are very common at the southern declivity of the Alps, between Lake Maggiore and Carinthia.

The *red sandstone* is generally composed of fragments of the rocks that exist in the neighbouring mountains. In the north of Germany, these fragments are more generally quartz, lydian stone, silex (hornstone), porphyry, syenite, and clay-slate, than gneiss, granite, and mica-slate. The colour of the red sandstone is very variable; it passes from reddish-brown to grey (graue liegende); these are sometimes mixed by very thin beds, as in the variegated sandstone (grès bigarré). The red tint of this formation is owing, according to the opinion of several celebrated geologists, to the ferruginous matter from the neighbouring porphyries. Without invalidating the correctness of this observation with respect to a part of the ancient continent, I wish to express some doubts respecting the influence of porphyries on the formation of the red sandstone in the equinoxial regions of the new continent. The sandstone of the vast steppes of Venezuela is reddish-brown, like the *todte liegende* of Mansfeld; it contains no fragments of porphyry, and no bed of intermediary or secondary porphyry is known for several hundred leagues. · It is the same with regard to the red sandstone of Fünfkirchen and Vasas in Hungary, described by M. Beudant.

Wherever, in the formation § 26., coarse conglomerates alternate with fine-grained arenaceous rocks, the latter pass to slaty coal-sandstone strongly micaceous (sandstein-schiefer). These alternating masses contain slate-clay which is greenish or brown, When this clay is strongly carburetted (kohlen-schiefer) and bituminous, it sometimes contains (Suhl, Goldlauter) argentiferous ores (grey copper, galena, and copper pyrites). It contains impressions of fossil fish, and has the aspect of the copper-slate belonging to the alpine limestone. On the other hand, the disintegration of fine-grained arenaceous rocks forms layers of quartzose and brownish sand (triebsand) in the middle of the most compact red sandstone (Walkenried and Bieber). The cement of the coal sandstone is sometimes limestone, and the portions of carbonate of lime become sometimes so frequent that they give to the rock the appearance of a granular and arenaceous limestone (carboniferous mountains on the limits of Hungary and Galicia). These are the *calcariferous sandstones* of M. Beudant, mixed with green chloritic grains. With respect to the fragments imbedded in the red sandstone, they are either angular and completely fixed in the mass, or rounded and flattened like the pebbles of the most recent nagelfluhe. The formation of the red sandstone which constitutes the greater part of Ireland, and which is so common in the north of Germany, in the Black Forest, and the Vosges, is almost entirely wanting (like the formation of porphyries) in the high Alps of Switzerland. The Niesen probably belongs to grauwacke, and M. Gruner believes that the vicinity of Mels, Bregentz, and Sonthofen, furnish the only conglomerates which by their structure and position resemble the red sandstone. In the high Alps, as well as in several parts of Silesia (Schweidnitz) and Hungary (Dunajitz), the red sandstone contains imbedded alpine limestone, with which it alternates; in the circle of Neuštadt in Saxony, the red sandstone is entirely wanting.

The beds subordinate to the red sandstone, or alternating with it, are the following; fetid limestone and strongly carburetted and bituminous slates, (kohlen-schiefer of Freiesleben), which denote the close connection of the red sandstone with zechstein and marno-bituminous slate (kupferschiefer); greenstone, a mixture of feldspar and hornblende (Noyant and Figeac, in France), sometimes even with pyroxene (Scotland); cellular mandelstein, sometimes extremely porous, containing (Ihlefield in the Hartz; banks of la Nahe, Oberstein, and Kirn; Heavitree, Exeter) agates, calcedony, prehnite, and chabasie, and penetrating, as by fissures, into the mass of the red sandstone (Planitz in Saxony); coal alternating with slate-clay with impressions of ferns; anthracite (Schönfeld, between Altenberg and Zinnwald) belonging more particularly, according to M. Beudant, to the porphyry interposed in the red sandstone, than to the latter rock; porphyries, first alternating with the red sandstone, and then covering them in great rocky masses; pechstein (quartz resinite, or retinite). The true position of pechstein in Saxony has been observed by MM. Jameson, Raumer, Przystanowsky, and Schenk. This substance forms a porphyry with a semi-vitreous base, containing feldspar often shivered, and very little mica, hornblende, and crystallised quartz (valley of Triebitch). The pitchstone contains imbedded fragments of gneiss (Mohorn and Braunsdorf); it is traversed by small veins of fibrous anthracite (Planiz, near Zwickau) and alternates with common porphyry of the red sandstone. These porphyries and retinites repose (Nieder-Garsebach) on transition syenite. M. Beudant, who has recently given a detailed description of this position, has observed that the pitchstone of Herzogswalde is contained in an arenaceous deposite with a basis of argilolite (thonstein). This deposite contains imbedded angular fragments of gneiss and mica-slate, and belongs to the red sandstone. The pitchstone of Grantola, at the lake Maggiore, is found in the same position; that of Scotland contains naphtha. In Peru there is pitchstone (smoke-grey, almost without feldspar, and containing crystallised mica), in the road from Couzco to Guamanga, where it forms mountains; but, according to the observations of M. de Nordenflycht, this formation is subordinate, as in Europe, to porphyritic rocks.

The whole of the formation § 26., which we now describe, is generally characterised by the absence of fossil shells. If a few are found, they belong to beds of limestone and carburetted schist (kohlenschiefer) which are interposed in the red sandstone, and not to the mass of this formation, which, in the two hemispheres, contains in abundance only trunks of fossil wood, and other remains of monocotyledon plants (plains of Thuringia, Kiffhauser, Tilleda; plains of Venezuela between Calabozo and Chaguaramas; table-land of Cuença, at the south of Quito). M. Adolphe Brongniart thinks, however, that the impressions of true palms are not found in this coal.

I had the opportunity of observing the formation of red sandstone in the equinoxial region of the new continent, north and south of the equator, in six different places; in New Spain (from 1100 to 1300 toises high), in the steppes or Llanos of Venezuela (30—50 toises) in New Grenada (50—1800 toises), on the southern table-land of the province of Quito (1350—1600 toises), and in the western valley of the Amazon (200 toises).

1°. *New Spain.* The schists and transition porphyries of Guanaxuato (table-land of Anahuac), of which we have given a detailed description above, (§§ 22, 23.) are covered with a formation of red sandstone. This formation fills the plains of Celaya, Salamanca, and Burras (900 toises); it there supports limestone very analogous to that of Jura, and lamellar gypsum. It extends by Cañada de Marfil, to the mountains that surround the town of Guanaxuato, and appears in insulated spots in the Sierra de Santa Rosa, near Villalpando (1330 toises). This Mexican sandstone has the most striking resemblance with the *rothe todte liegende* of Mansfeld in Saxony. It contains angular fragments of lydian stone, syenite, porphyry, quartz, and flint (splittriger hornstein). The cement that unites these fragments is argilo-ferruginous, very tenacious, yellowish-brown, and often (near the river of Serena) of a brick-red colour. Beds of coarse conglomerate, containing fragments two or three inches in diameter, alternate with a fine-grained conglomerate, sometimes even (Cuevas) with a sandstone consisting uniformly of grains of quartz. Coarse conglomerates abound more in plains and ravines than on the heights. I thought I perceived in the most ancient beds (mine of Rayas), a passage from the red sandstone to grauwacke; the pieces of imbedded syenite and porphyry become very small; their outlines are indistinct and appear as if softened into the mass. This conglomerate (frijolillo de Rayas) must not be confounded with that of the mine of Animas, which is whitish-grey, and contains fragments of compact limestone. In the red sandstone of Guanaxuato, as well as in that of Eisleben in Saxony, the cement is often so abundant (road from Guanaxuato to Rayas and Salgado), that the imbedded fragments are no longer distinguished. Argillaceous beds, from three to four inches thick, then alternate with the coarse conglomerate. The great formation of red sandstone, superposed on metalliferous clay-slate, appears in general only when supported by transition porphyry (Belgrade, Buffa de Guanaxuato); but we see it distinctly placed on the latter rock at Villalpando. I found no petrified shells, nor any trace of coal or fossil wood in the red sandstone of Guanaxuato. These combustible substances occur frequently in other parts of New Spain, especially in those which are least elevated above the level of the sea. In the interior of New Mexico, coal is known not far from the banks of the Rio del Norte; other deposites of it probably are hid in the plains of Nuevo-Sant-Ander and the Texas. At the north of Natchitoches, near the coal-mine of Chicha, subterraneous detonations are heard from time to time, from an insulated hill, occasioned perhaps by the inflammation of hydrogen gas mixed with atmospheric air. Fossil wood is common in the red sandstone that extends towards the north-east of the town of Mexico. It is also found in the immense plains of the Intendance of San-Luis-Potosi, and near the town of Altamira. The coal of Durasno (between Tierra-Nueva, and San-Luis de la Paz) is placed below a bed of clay containing fossil wood, and over a bed of sulphuret of mercury which covers the porphyry. Does it belong to very recent lignites? or, ought we not rather to admit that these combustible substances of Durasno, these clays and semi-vitreous porphyries (pechstein-porphyre), globular and covered with mammillated hyalite, porphyries which in other parts of Mexico (San Juan de la Chica; Cerro del Fraile near the Villa of San Felipe) contain deposites of sulphuret of mercury, are connected with the great formation of red sandstone? There is no doubt that this formation is as rich in mercury in the new continent as in the west of Germany; it is found there when the porphyries are wanting (Cuença, table-land of Quito); and if the union of veins of tin with veins of cinnabar, in the porphyries of San Felipe, appear at first to remove porphyritic rocks which abound in mercury from those of the red sandstone, we must recollect that transition clay-slate and porphyries (Hollgrund near Steben, Hartenstein) are also sometimes stanniferous in Europe.

I place in the suite of coal-sandstones of Guanaxuato, a formation that is somewhat doubtful, which I have already described in my *Political*

Essay on New Spain, by the name of *lozero*, or feldspathic conglomerate; it is an arenaceous rock, reddish-white, and sometimes apple-green, which divides, like sandstone (*Leuben* or *Wald-plattenstein* of Suhl), in very thin plates (*lozas*); it contains grains of quartz, small fragments of clay-slate, and a quantity of crystals of feldspar partly broken and partly entire. These various substances are connected together in the *lozero* of Mexico, as in the rock of porphyritic aspect of Suhl, by an argillo-ferruginous cement (Cañada de Serena, and almost the whole mountain of that name). It is probable, that the destruction of the porphyry has had great influence on the formation of the feldspathic sandstone of Guanaxuato. The most experienced mineralogist might at first be led to take it for a porphyry with an argillaceous base, or for a porphyritic breccia. Around Valenciana the *lozero* forms masses of 200 toises in thickness, and which exceed in elevation the mountains formed by the intermediary porphyry. Near to Villalpando a feldspathic conglomerate, with very small grains, alternates by beds one or two feet thick, twenty-eight times with slate-clay of a darkish-brown. I saw every where this conglomerate or *lozero* reposing on the red sandstone, and at the south-west declivity of Cerro de Serena, in descending towards the mine of Rayas, it appeared to me sufficiently evident that the *lozero* forms a bed in the coarse conglomerate of Marfil. I doubt, consequently, if this remarkable formation can belong to trachytic pumice-conglomerates, as M. Beudant seems to think from its analogy with some rocks in Hungary. The argillaceous cement becomes often so abundant that the imbedded parts are scarcely visible, and the mass passes to compact claystone (thonstein). In this state the *lozero* furnishes the fine building-stone of Queretaro, (quarries of Caretas and Guimilpa,) which is so much esteemed for construction. I have seen columns fourteen feet high, and two feet and a half in diameter, flesh-red, brick-red, or peach colour. When in contact with the atmosphere those fine colours change to grey, probably by the action of the air on the dendritic manganese contained in the fissures of the rock. The columns of Queretaro have a smooth fracture, like that of the lithographic Jura limestone. With difficulty we discover in these claystones (argilolites) some very small fragments of clay-slate, quartz, feldspar, and mica. I will not decide, if the unbroken crystals of *lozero* or feldspathic sandstone are developed in the mass itself, or are found there accidentally. I shall here confine myself to the observation, that in Europe this red sandstone and these porphyries are also sometimes characterised by a *local suppression* of crystals and imbedded fragments. The *lozero* appears to me to be a formation of superposed sandstone, perhaps even subordinate to the red sandstone; and if no rock entirely similar is found in the ancient continent, we see, at least, the first germ of this kind of pseudo-porphyritic structure in the layers of sandstone with feldspar crystals, broken or entire, which are sometimes imbedded in the great formation of red sandstone of Mansfeld and Thuringerwald. (Freiesleben, *Kupf.*, b. iv. p. 82. 85. 95. 194.).

2°. *Venezuela.* The immense plains of Venezuela in South America (Llanos of Lower Oroonoko) are, for the most part, covered with red sandstone, limestone, and gypsum. The red sandstone is there disposed in a *concave position* (muldenförmige Lagerung) between the mountains of the shore of Caracas and those of Parima, or the Upper Oroonoko. It is connected at the north with transition slate, and at the south reposes immediately on primitive granite. It is a conglomerate of rounded fragments of quartz, lydian stone, and kiesel-schiefer, united by an argillaceous and ferruginous cement that is olive-brown and extremely tenacious. This cement is sometimes (near Calabozo) of so vivid a red, that the people of the country imagine it is mixed with cinnabar. The coarse conglomerate alternates with a fine-grained quartzose sandstone (Mesa de Paja). Small masses of brown iron, and of the petrified wood of monocotyledon plants are imbedded in both. This arenaceous formation is covered (Tisnao) by a whitish-grey compact limestone, analogous to that of Jura. Above this limestone we find (Mesa de San Diego and Ortiz) lamellar gypsum alternating with beds of marl. I saw no fossil shells in any of these beds, whether arenaceous, calcareous, gypseous, or marley. The cement of the conglomerate does not effervesce any where with acids, and the sandstone of the steppes of Venezuela appeared to me to be far removed by its place and composition from the *nagelfluhe* (sandstone with lignite) of the tertiary formation, with which it has some analogy of aspect by the rounded form of the imbedded fragments. These arenaceous and limestone formations do not rise above thirty or forty toises of absolute height. In the eastern part of the Llano of Venezuela (near Curataquiche) fine pieces of ribband-jasper or Egyptian *pebbles* are dispersed on the surface of the soil. Do they belong to the red sandstone, or, as near Suez, to a more modern formation?

3°. *New Grenada.* A formation of sandstone of immense extent covers almost without interruption, not only the northern plains of New Grenada between Mompox, the canal of Mahates, and the mountains of Tolu and Maria, but also the basin of the Rio de la Magdalena (between Teneriffe and Melgar), and that of Rio Cauca (between Carthago and Cali). Some scattered fragments of slaty and coal-sandstone (kohlen-schiefer), which I found at the mouth of the Rio Sinu (at the east of the Gulf of Darien), render it probable that this formation extends even towards the Rio Atrato and the Isthmus of Panama. It rises to a great height, not on the intermediary or central branch of the Cordillera (Nevados de Tolima and de Quindiu), but on the eastern branch (Paramos de Chingasa and de Suma Paz), and on the western (mountains between the basin of the Rio Cauca and the platiniferous formation of Choco). I traced this sandstone of New Grenada, without losing sight of it, from the valley of Rio Magdalena (Honda, Melgar, 130—188 toises), by Pandi, as far as the table-land of Santa Fé de Bogota (1365 toises), and even above the lake of Guatavita, and the chapel of Our Lady of Montserrate. It leans on the eastern Cordillera (that which separates the tributary streams of Rio Magdalena from those of Meta and the Oroonoko), at more than 1800 toises of height above the level of the ocean. I dwell on these notions of mineralogical geography, because they furnish new proofs of the enormous thickness which rocks attain in the equinoxial regions of America. Several secondary formations (sandstone with beds of coal, gypsum with rock-salt, limestone almost without petrifactions), which, in the table-land of Santa Fé de Bogota, we should be tempted to take for a group of local formations filling a basin, descend as far as the valleys, of which the level is 7000 feet lower than this table-land. In going from Honda to Santa Fé de Bogota, the sandstone is interrupted near Villeta by transition clay-slate; but the position of the salt-springs of Pinceima and Pizara near Muzo leads me to think, that on that side also, on the banks of the Rio Negro (between the amphibolic and carburetted slates of Muzo containing emeralds, and the transition slates with copper veins of Villeta), the coal-sandstone and muriatiferous gypsum of the table-land of Bogota and Zipaquira are connected with the homonymous formations which fill the basin of Rio Magdalena, between Honda and the streight of Carare.

This sandstone of New Grenada (where I was able to examine it between 4° and 9° of north latitude,) is composed of alternate beds of fine-grained quartzose and slaty sandstone, and conglomerates that contain angular fragments (from two to three inches thick) of lydian stone, clay-slate, gneiss, and quartz (Honda, Espinal). The cement is argillaceous, ferruginous, and sometimes siliceous. The colours of the rock vary from yellowish-grey to brownish-red: the latter colour is owing to iron; a brown ironstone is every where found, very compact, occurring in nests, small beds, and irregular veins. The sandstone is stratified in layers more or less horizontal: these layers are sometimes inclined in groups always in the same manner. Near Zambrano, on the western bank of Rio Magdalena, south of Teneriffe, the rock assumes a globular structure. I have seen balls of small-grained sandstone of two or three feet in diameter; they separate into twelve or fifteen concentric layers. Lydian stone of the finest black, rarely traversed by veins of quartz, is much more abundant in the coarse conglomerates than fragments of primitive rocks. Every where slaty fine-grained sandstone occurs in greater masses than the conglomerates with large fragments. The latter disappear almost entirely on the heights (above 800 to 1000 toises). The sandstone of the table-land of Bogota and that which is observed in ascending to the two chapels placed above the town of Santa Fé, at 1650 to 1687 toises high, are uniformly composed of small quartzose grains. Scarcely any longer can fragments of lydian stone be observed; the grains of quartz are so close together that the rock sometimes assumes the aspect of a granular quartz. It is this very quartzose sandstone that forms the natural bridge of Icononzo. These arenaceous rocks no where effervesce with acids. Besides brown iron-stone, and (what is more extraordinary) some nodules of very pure graphite, this formation also contains, at every height, beds of a brown clay, soft to the touch, and not micaceous. This clay (Gachansipa, Chaleche, mountain of Suba,) sometimes becomes strongly carburetted, and passes to bituminous slate. The aperient salt of Honda (sulphate of magnesia), so celebrated in those countries, appears as an efflorescence on those argillaceous beds (Mesa de Palacios near Honda). The sandstone no where displays varied colours disposed in zones, nor those insulated masses of clay of a lenticular form which characterise the *variegated sandstone* (bunte sandstein), that is, the sandstone that covers alpine limestone, or zechstein. I saw the formation of sandstone we have just described, reposing immediately on a granite with tourmalines (Peñon de Rosa at the north of Banco, valley of the Magdalena; cascade of la Peña near Mariquita), on gneiss (Rio Lumbi, near the abandoned mines of St. Anne), and on transition clay-slate (between Alto de Gascas and

Alto del Roble, north-west of Santa Fé de Bogota). We know no other secondary rock below the sandstone of New Grenada. It contains caverns (Facatativa, Pandi), and affords considerable beds, not of lignite, but of lamellar and compact coal mixed with jet (pechkohle), between la Palma and Guaduas (600 toises), near Velez and the Villa de Leiva, as also in the table-land of Bogota (Chipo near Canoas, Suba; Cerro de los Tunjos), at the great height of 1370 toises. Remains of organised bodies of the animal kingdom are extremely rare in this sandstone. I only once found trochilites (?) almost microscopic in an interposed bed of clay (Cerro del Portachuelo, at the south of Icononzo). These coals of Guaduas and Canoas might have been a more recent formation, superposed on the red sandstone, but nothing appeared to me to prove this superposition. This piciform coal (jayet, pechkohle) belongs, no doubt, rather to the lignites of the tertiary sandstone and the basalts; but it also certainly forms small beds in the slaty coal (schieferkohle), of the porphyry and red sandstone formation.

The formations which cover the sandstone of New Grenada, and which characterise it, I believe, more particularly as the red sandstone in the series of secondary rocks, are fetid limestone (confluence of Caño Morocoy and the Rio Magdalena), and lamellar gypsum (basins of Rio Cauca near Cali, and of Rio Bogota near Santa Fé). In the two basins of Cauca and Bogota, of which the height differs nearly 900 toises, we see the three formations of coal-sandstone, gypsum, and compact limestone, succeed each other very regularly from below upwards. The two latter seem to constitute but one formation, which represents alpine limestone or zechstein, and which, though generally without petrifactions, contains some ammonites at Tocayma (valley of Rio Magdalena). Gypsum is often wanting, but at the great elevation of 1400 toises (Zipaquira, Enemocon, and Sesquiler) it is muriatiferous, furnishing deposites (salzthon) of rock-salt mixed with clay, which have been worked for ages on a great scale.

From the whole of the observations which I have just stated on the position of the sandstone of New Grenada, I do not hesitate to regard that rock, which, in its developement is 5000 or 6000 feet thick, and which will soon be again examined by two learned travellers, MM. Boussingault and Rivero, as a red sandstone (todtes liegende), and not as a variegated sandstone (sandstones of Nebra). I am not ignorant that frequent beds of clay and brown ironstone belong more particularly to the variegated sandstone (grès bigarré), and that the oolites are also often wanting in this sandstone. I know also, that in Europe the variegated sandstone (placed above zechstein) presents some traces of coal, small beds of very quartzose sandstone (granular quartz), and rock-salt, and that this latter substance belongs to it even exclusively in England. All these analogies would appear to me highly important, if beds of coarse conglomerate alternating (in low regions) with beds of small-grained sandstone, if angular fragments of lydian stone, and even of gneiss and mica-slate imbedded in coarse conglomerates, did not characterise the sandstone of New Grenada as parallel to the red sandstone or coal-sandstone; that is, as being parallel to that which immediately supports alpine limestone (zechstein) containing gypsum and rock-salt. When the variegated sandstone (north of England and Wimmelburg in Saxony) sometimes exhibits fragments of granite and syenite, these fragments are rounded, and simply enveloped with clay; they do not form a compact and tenaceous conglomerate with angular fragments, like the red sandstone. The latter rock abounds in Mansfeld as well as in New Grenada, in interposed masses of clay (Cresfeld, Eisleben, Rothenberg), and in small beds of brown and red iron-ore (Burgörner, Hettstedt). The globular structure of the sandstone of the Rio Magdalena is found also in the coal-sandstone of Hungary (Klausenburg), in the whitish conglomerate of Saxony (weiss-liegendes de Helbra), which connects the coal-sandstone with zechstein, and according to the observations which I made with M. Freiesleben, in 1795, even near Lausanne, in the molasse of Argovia (tertiary sandstone and lignite). It is the whole of these relations of position that determines the age of a formation, and not its composition and structure only; geognosts who have become acquainted with the different formations of sandstone, not from cabinet specimens, but from frequent excursions in the mountains, are well aware that if (by the suppression of alpine limestone, muschelkalk, limestone of Jura, and chalk) the red sandstone, the variegated sandstone mixed with clay, quadersandstein which is not always white and quartzose, and molasse alternating with coarse pudding-stone (nagelfluhe), were immediately superposed on each other, it would be difficult to decide on the limits of these four arenaceous formations, of an age so different.

The red sandstone of New Grenada appears to dip, in the northern part of the basin of Rio Magdalena (between Mahates, Turbaco, and the coast of the sea of the Antilles), beneath a tertiary limestone filled with madrepores and marine shells, and constituting, near the port of Carthagena, the Cerro de la Popa. But when we rise to the height of 1400 toises, the formation of limestone and gypsum supported by the red sandstone, is covered (Campo de Gigantes, at the west of Suacha in the basin of Bogota) by alluvial deposites, in which I found enormous bones of the mastodon. From the too general tendency, perhaps, of the modern geognost, to extend the domain of intermediary and tertiary formations, at the expense of the secondary formations, we might be tempted to regard the sandstone of Honda, the gypsum with rock-salt of Zipaquira, and the limestone of Tocayma and Bogota, as formations posterior to the chalk. According to this hypothesis, the coal of Guaduas and Canoas would become lignites, and the rock-salt of Zipaquira, Enemocon, Sesquiler, and Chamesa, entirely without vegetable remains, would be a formation parallel to the saliferous deposites (with lignite) of Galicia and Hungary, which, in the opinion of M. Beudant, belong to the tertiary rocks. But the aspect of the country, the almost total of want of organised fossils, observed as far as 10,000 feet of perpendicular height, the magnitude of the arenaceous and limestone beds uniformly extended, having no siliceous nodules, and infiltrations very compact, and no where mixed with sand and other incoherent matter, is opposed to those ideas, I had almost said to those encroachments of tertiary formations on secondary formations. The whole of the phenomena I have stated, leads me to believe that the sandstone of New Grenada, containing fragments of lydian stone and primitive rocks, is the true red sandstone of the ancient continent. We do not know whether this sandstone, which I saw at 1700 toises of height at the western declivity of the Cordillera of Chingasa (a Cordillera that separates the town of Santa Fé de Bogota from the plains of Meta), reaches the summit of that great chain of mountains, in stretching towards the plains of Casanare. It may be conjectured; for the deposites of rock-salt, and the springs of muriate of soda, succeed each other in traversing the eastern Cordillera of New Grenada, from Pinceima as far as Llanos du Meta (by Zipaquira, Enemocon Tausa, Sesquiler, Gachita, Medina, Chita, Chamesa, and El Receptor), from south-west to north-east, in the same direction, on a distance of more than fifty leagues. In every region of the globe this disposition of salt-springs by bands (or fissures?) more or less prolonged is observed. When you advance towards the Oroonoko from the saliferous plains of Casanare, the secondary formations disappear by degrees, and in the Sierra Parime, granite-gneiss every where is seen. On the banks of the Oroonoko only, near the great cataracts of Atures and Maypures, small fragments of ancient conglomerate superposed on primitive rocks are found. This conglomerate contains grains of quartz and (Isla del Guachaco) even fragments of feldspar united by a cement which is olive-brown, argillaceous, and very compact. This cement, where it is abundant, exhibits a conchoidal fracture, and passes to jasper. This arenaceous rock, which, I believe, belongs to the red sandstone of the steppes of Venezuela, contains some very flat masses of brown iron ore, and reminds us of those sandstones which in Upper Egypt and Nubia, repose also immediately on the gneiss-granite, gneiss of the cataracts of the Nile.

4°. *Table-land of Quito*. In the ancient hemisphere the Cordilleras of Quito exhibit the most extensive formation of red sandstone which I have hitherto observed. That rock, from 1300 to 1500 toises high above the level of the sea, covers, on a length of twenty-five leagues, the whole table-land of Tarqui and Cuença; which is become celebrated by the operations of the French astronomers. It rises in the Paramo de Saar as far as 1900 toises, and the thickness of the whole mass exceeds 800 toises. It reposes at the north (Cañar, southern declivity of Assuay) and south (Alto de Pulla near Loxa) on micaceous primitive slate. The formation of red sandstone, in the province of Quito, is coloured by brown and yellow iron ore, of which it contains numerous veins. The sandstone is generally very argillaceous, with small grains of quartz a little rounded; but it is also sometimes slaty, and alternates, as in Thuringia, with a conglomerate that contains fragments of porphyry, of three, five, and even nine inches in diameter. We find in this formation, beds of clay, sometimes brown (Tambo de Burgay and banks of Vinayacu), sometimes white and steatitic, passing to the clay porphyries (thonstein) of the red sandstone (Rio Uduchapa and Cerro de Coxitambo), and covered, in contact with the atmospheric air, with nitrate of potass (Cumbe); trunks of the petrified wood of monocotyledons (ravine of Silcayacu, where I saw pieces four feet long and fourteen inches thick); mineral pitch, both fluid and hardened into asphaltum, with a conchoïdal fracture (Parĉhe and Coxitambo); flint (splittriger hornstein) passing to silex pyromaque or agathe (Delay); veins of sulphuret of mercury (Cerros de Guazun, and Upar north-east of the village of Azogues); beds of pulverulent black oxid of manganese (at the west of the town of

Cuença); granular and lamellar limestone (Portete, at the western bank of Llano de Tarqui). This limestone formation, called, very improperly, in that country, riband-jasper, exhibits alternate beds of opake and saccharoïd limestone, similar to the marble of Carara, and of fibrous and undulated limestone, with milky streaks. The whole mass is diaphanous like the finest eastern alabaster (memphitic or phengite marble of the ancients). I should have been tempted to take this rock of Tarqui, which is sought after by sculptors, like the alabaster of Florence and the marble of Tolonta (between Chillo and Quito), for a variety of travertino or fresh-water formation, if it had not appeared to me at the south of Cuença, on the bank of the Rio Machangara (according to the inclination of the beds), to be interposed to the red sandstone which I have just described. We must, however, distinguish this translucid and striped marble of Tarqui from the granular and opake limestone of Cebollar, which appears a little north of Cuença, and which, covered by red sandstone, is probably (§ 10.) superposed on the mica-slate of Cañar. In the volcanic parts of the Andes, the table-lands or elevated basins are filled, some with secondary formations, covering transition porphyries; others with tertiary and fresh-water formations, superposed on trachytic tufas. When well-informed geognosts shall be resident in the great towns placed at the back of the Cordilleras, towns which will become the centre of American civilisation, they will be enabled to decide with certainty on those insulated portions of limestone, gypsum, and arenaceous rocks, that are found at a height of between 1200 and 1600 toises.

5°. *Peru.* The formation of red sandstone of Cuença, which is covered on several points with beds of lamellar gypsum (Muney, Juncay, and Chalcay, at the west of Nabon), is found repeated in Upper Peru, at the height of 1460 toises, in the great table-land of Caxamarca. The sandstone of Caxamarca is also argillaceous, without shells, and filled with brown iron ore. It appeared to me to be supported by porphyries of a trachytic aspect (Cerros de Aroma and Cundurcaga). It supports the alpine limestone of Montan and Micuipampa, celebrated for its metallic riches. The thermal hydrosulphurous waters that issue from the sandstone of Cuença (south lat. 2° 53′), and of Tollacpoma near Caxamarca (south lat. 7° 8′), have almost the same temperature, 72° and 69° cent.

The analogy observed between the red sandstone of New Grenada, Peru, and Quito, and the red sandstone of the country where Füchsel (*Historia terræ et maris ex historia Thuringiæ eruta*) gave the first description of the great coal formation, must strike experienced geognosts. I shall not dwell on the phenomena so well known, of the alternation of coarse conglomerates, and fine-grained sandstones; nor on the absence of fragments of limestone, of which we find only one very rare example in the pudding-stones of the red sandstone in the Pyrenees (valley of Barillos); nor of the interposed beds of coal, clay, brown iron ore, and limestone: I shall confine myself to the observation, that the red sandstone of Germany contains mercury (Mörsfeld and Moschellandsberg in the duchy of Deux-ponts, as at Dombrava in Hungary); the petrified wood of monocotyledon plants (Siebigkerode, Kelbra, and Rothenburg, in Thuringia); agates, chert and common flint (hornstein and feuerstein) passing to calcedony (Kiffhäuer, Wiederstädt, Goldlauter, and Grossreina, in Saxony, in the coarse conglomerate of the red sandstone; Oberkirchen and Tholey in the duchy of Deux-ponts; Netzberg near Ilefeld, at Hartz, in the mandelstein of the red sandstone); and mineral bitumen (Naundorf and Gnölzig in the county of Mansfeld). All these phenomena occur also in the part of equinoxial America through which I passed.

6°. *Banks of the Amazon.* The great basin of the river of the Amazons exhibits, at least in the eastern part, the same phenomena which we have pointed out in giving the geognostic description of the Llanos of Venezuela, and the basin of the Oroonoko. In descending from the summit of the granitic Andes of Loxa by Guancabamba to the banks of Chamaya, we find a sandstone with an argillaceous cement, superposed on the transition porphyries of Sonanga, and covered (between Sonanga and Guanca) with a limestone containing gypsum and rock-salt. The sandstone of Chamaya fills the plains of Jaen de Bracamoros, at 190 to

260 toises of height above the level of the ocean. It forms hills with abrupt declivities, resembling fortifications in ruins. We there distinguish beds with small rounded grains of quartz, and coarse conglomerates composed of pebbles of porphyry lydian stone and quartz, of two or three inches in diameter. The coarse conglomerates are somewhat rare; they form, however, the *pongo* of Rentema, and other rocky dykes that cross the Upper Maragnon, and impede the navigation of the river. I could never discover among the imbedded fragments in the sandstone of Chamaya any one of a limestone rock. This circumstance, together with the presence of lydian stone mixed through the mass, the alternation of grey fine-grained sandstone and coarse conglomerates, every where so rare in the variegated sandstone (Schochwitz in Saxony); finally, the superposition of zechstein and gypsum with rock-salt to the sandstone of the Amazon, lead me to admit the identity of that formation and those of Cuença and Caxamarca, notwithstanding the difference of more than 1000 toises of absolute height. We have already seen in New Grenada, coal sandstone descend from the great table-land of Bogota to the plains of the Rio Magdalena. One very remarkable peculiarity, and which seems at first to separate the sandstone of the Amazon and the Chamaya from that of Europe, is the intercalation of some beds of sand in parts altogether disintegrated. I saw between Chamaya and Tomependa, layers of quartzose sandstone, three or four feet

thick, alternating with beds of siliceous sand from seven to eight feet. The parallelism of these beds, which are not much inclined, extends to great distances. I am aware that the mixture of sand and solid sandstone characterises more particularly the variegated sandstone; that which covers zechstein (Wimmelburg and Cresfeld in Saxony), and the tertiary sandstone above the gypsum with bones (Fontainebleau near Paris); but MM. Voigt and Jordan found also layers of sand (triebsand) in the red or coal sandstone (Röhrig near Bieber, and Kupferberg near Walkenried). We might believe that the analogy we have just remarked between the marine sandstone and sand of the tertiary formation, is strengthened to a certain degree by the frequency of petrified echini which we saw scattered on the surface of the soil on the beach of the Amazon, at 195 toises, and near Micuipampa, at more than 1800 toises high; but it may happen, that in regions hitherto so little examined, very new limestone formations repose on zechstein, and nothing seems to denote that the sandstone of Chamaya, alternating at the same time with beds of sand, and conglomerates with fragments of porphyry and lydian stone, is a tertiary sandstone similar to that of the Parisian formations.

I ought, perhaps, to place the zechstein or alpine limestone immediately after the coal-sandstone, because those two rocks sometimes constitute but one formation; but I prefer describing first the quartz formation of Guangamarca (flözquartz), as being parallel to the coal sandstone. It is a *geognostic equivalent*, peculiar to the southern hemisphere.

SECONDARY QUARTZ-ROCK.

§ 27. This remarkable formation, entirely unknown to the geognosts of Europe, predominates in the Andes of Peru, between the 7° and 8° of southern latitude. I have seen it reposing indifferently on transition porphyries (at the eastern declivity of the Cordillera, Cerro of N. S. del Carmen, near San Felipe, 982 toises; Paramo de Yanaguanga between Micuipampa and Caxamarca, 1900 toises; at the western declivity of the Cordilleras, Namas and Magdalena, 690 toises); and on primitive granite (Chala, near the coast of the Pacific Ocean, 212 toises). This superposition on rocks of a very different age proves the *independence* of the formation which we describe. It is much less developed at the eastern than at the western declivity of the Andes. At the latter it attains a thickness of several thousand feet, reckoning perpendicularly to the planes of stratification; it there replaces the red sandstone, supporting immediately (Indian villages of la Magdalena and Contumaza) zechstein or alpine limestone. It is either the latest of the transition formations, or the most ancient of the secondary formations; it is a real compact or granular quartz, not porous or cellular, most frequently greyish-white, or yellowish and opaque, and not mixed either with talc or mica.

This formation is sometimes compact and with a scaly fracture, like quartz in beds (lagerquartz of primitive gneiss-granite); sometimes with very fine grains, similar to that of the quartz of the transition limestone of the Tarantaise. It is consequently neither an arenaceous rock, nor a variety of the quartzose sandstone with a siliciferous cement, in which the cement disappears by degrees, and which belongs both to the variegated sandstone (Detmold), quadersandstein, green sandstone, plastic clay (trappsandstein), and to the sandstone of the tertiary formation (forest of Fontainebleau). The deep ravines that furrow the declivity of the Cordilleras, and the immense number of blocks torn from their natural position, facilitate the observation of this formation of quartz, which is very homogeneous, destitute of shells, and also of subordinate beds. I examined it for several days, expecting to find, in a rock covered by zechstein and replacing the red sandstone, some traces of a cement, of grains or agglutinated fragments: my researches were fruitless; I could no where convince myself that this compact or granular quartz was an arenaceous or fragmentary rock. It is sometimes very regularly separated into beds of eight inches to two feet thick, directed (Aroma, Magdalena, and Cascas) N. 53°, 68° W., and inclined from 70° to 80° S.E. At the eastern declivity of the Andes, on the banks of the Chamaya, a bed of quartz, similar to that which I have just described, appears interposed in a formation of greyish-blue compact limestone.

This limestone is not a transition rock (as might be thought from the position of the compact quartz of Pesay and Tines in the Tarantaise, § 20.); the number and nature of its shells, on the contrary, as well as the sinuosity of its beds, seem to bring it nearer to the zechstein or the alpine limestone. It is not extraordinary to see a siliceous rock which supports limestone penetrate into it, and there form an interposed bed. This circumstance also occurs, but in veins (Cerro de N. S. del Carmen near San Felipe), in the formation on which quartz rock reposes. The alpine limestone of San Felipe covers this rock, which is placed on green transition porphyry traversed by veins of quartz three feet thick.

It may be useful at the end of this article to observe, that we must not confound nine formations of quartz and quartzose sandstone of primitive, intermediary, secondary, and tertiary formation, of which the second and fourth only are *independent*, while the others form but subordinate beds: 1°, quartz (lager-quartz) belonging to granite-gneiss, mica-slate, and primitive clay-slate; 2°, chloritous, or talcose quartz of Minas Geraes, of Brazil and Tiocaxas, in the Andes of Quito; an independent primitive formation, succeeding to clay-slate (§ 16.), or replacing it as in Norway; 3°, compact transition quartz, described by MM. Brochant, Haussmann, and Leopold de Buch, and subordinate (§ 20.) to the limestone, and slaty rocks of the Tarantaise, Kemi-Elf in Sweden, and Skeen in Norway (§ 23.);

4°, secondary quartz (§ 27.) parallel to the red sandstone, and penetrating into the alpine limestone of the Andes of Contumaza and Huancavelica. We may join to these formations of quartz, masses entirely quartzose; 5°, of the variegated sandstone; 6°, of the quadersandstien; 7°, the green sandstone, or secondary sandstone with lignite, placed between the Jura limestone, and the chalk; 8°, the sandstone belonging to tertiary sandstone, with lignite (plastic clay) above the chalk; 9°, the sandstone of Fontainebleau. A rock is determined with more certainty, when we have before us the table of formations which are analogous in their composition, but very different in their position.

II. ZECHSTEIN OR ALPINE LIMESTONE (MAGNESIAN LIMESTONE); HYDRATED GYPSUM; ROCK-SALT.

§ 28. The word *zechstein* is usually applied by the miners and geognosts of Germany to a part of the formation which we are about to describe; they distinguish compact limestone (zechstein) from the copper-slate which it immediately covers, and the superposed gypsum and fetid limestone. I call by the term zechstein the entire group, of which that rock is the geognostic representative. It is a great limestone formation, that immediately succeeds to the red sandstone, or coal-sandstone, and with which it is sometimes so closely connected, that they are found interstratified. The upper limit of the zechstein is more difficult to fix; in

Germany, and in several parts of the east of France, this rock terminates where the variegated sandstone, or the sandstone of the oolites begins (bunte sandstein). In England, magnesian limestone, representing zechstein by its position, is covered with a marly and muriatiferous formation (red marl), which exhibits a great analogy with the variegated sandstone of Germany, for in the latter we find more beds of clay, and marl of true sandstone. As, on the other hand, the rock-salt of England belongs to the red marl, while the rock-salt of the greater part of the continent belongs to the zechstein, we may admit that, of the two formations, which are nearly parallel, of red marl and variegated sandstone, containing marl, clays, and oolites, the first is more closely connected with the zechstein, and the latter with the muschelkalk, and, when this and quadersandstein are not developed, with the marly and oolitic limestone of Jura. It is perhaps from similar inductions, that Mr. Buckland, in his excellent table of the formations of England, published in 1816, united the magnesian limestone and the red marl or new red sandstone, into the same formation. However great may be the importance which we attach to geognostic affinities, as well as to the phenomena of alternation and penetration, observed in rocks immediately succeeding each other, we do not think ourselves less authorised in separating the various formations of the red sandstone, zechstein, and variegated sandstone, where we have seen them take an extraordinary developement in the two hemispheres.

In the course of this essay, I have often, after the example of many celebrated geognosts, made use of the more agreeable term of alpine limestone to designate zechstein; although I am well aware that, according to the excellent observations of MM. de Buch and Escher, the greater part of the limestones that constitute the high Alps of Switzerland are transition limestones (§ 22.). At a period when geognosy has been so much obscured from the creation of vague denominations, which are adopted only by a small number of mineralogists, I determined to change nothing in the received nomenclature, however vicious or barbarous it might appear. The imperfections of the language of geognosts are only dangerous to science when the position of every formation, and the limits within which those formations are found circumscribed, are not defined with sufficient clearness. In South Bavaria, in the Tyrol, Styria, and the country of Salzbourg, the high Alps of Benedictbaiern, Chiemsée, Hall, Ischel, Gmünden, and Untersberg, are very probably of zechstein. That rock, at Montperdu in the chain of the Pyrenees, mixed with fetid limestone, rises to a height of more than 1750 toises. Zechstein, in the Andes of Peru, very distinct from transition limestone, contains petrified shells on the ridge of the mountains between Guambos and Montan, and near Micuipampa (1400—2000 toises); and between Yauricocha and Pasco (2100 toises); near Huancavelica, Acoria, and Acobamba (2100—2207 toises). We see by these examples, that zechstein attains a very great elevation, north and south of the equator. It is certainly found in the alpine region of the Pyrenees, the Tyrol, and the Andes; but the term *alpine limestone* does not indicate that all the limestone alps in both hemispheres are composed of zechstein, any more than the word *coal-sandstone* denotes that coal belongs solely to the red sandstone. The question, which of the alpine summits of Switzerland and the Tyrol are of zechstein, and what summits are of transition limestone, is rather a question of mineralogical geography than a problem of general geognosy. The science of *formations* is confined to the description of a rock placed in the series of secondary formations, between coal-sandstone and the variegated sandstone alternating with clay; it does not decide on the great number of rocks, the position of which displays no certain diagnostic character; for instance, on limestone rocks not covered, and placed immediately on mica-slate or grauwacke. Wherever the coal-sandstone is wanting we cannot judge of the age of the limestone rocks, but from analogies of composition and interposed beds; we connect them with certain groups, as the botanist places, temporarily, in a certain well known genus, a plant of which he has not yet examined the fruit. Those doubts, far from proving the uncertainty of classifications are rather in favour of the method which should be pursued in positive geognosy.

Zechstein, considered in its most general point

of view, is sometimes (in the loftiest mountains) a simple formation, and sometimes (in the plains) it is composed of several small partial formations, which alternate together (Thuringia, Figeac, Autun, Villefranche). Its colour, which is most frequently greyish and bluish, is sometimes reddish; it passes, particularly in high regions, from compact to fine-grained granular limestone, and in that case it is traversed by small veins of calc-spar. These characters of colour and fracture are not, however, of great importance, for, according as the colouring matter (carburet of hydrogen and iron) is variously diffused, zechstein and transition limestone sometimes have the same colours; the former becoming blackish, and the latter greyish-white. Thus the black colour is found (duchy of Anhalt-Dessau, Hettstädt, Osnabrück) as far as in the muschelkalk. M. Freiesleben has well observed that zechstein is not generally dull, but somewhat brilliant (schimmernd) on account of the intimate mixture of small plates of calc-spar. This lustre, much less, no doubt, than in transition limestone, is remarked not only in very elevated mountains, but even in the zechstein of the plains. There also this rock becomes sometimes fine-grained granular (at Dester and near Hameln; between Bolkenhayn and Waldenbourg, and near Tarnowiz in Silesia). I found the same tendency to a crystalline structure, in the zechstein of Mexico, and in that of the Llanos of Venezuela. It is not occasioned, as in the Jura limestone, by numerous fragments of organic remains, and it would be wrong to attribute this tendency exclusively to the transition limestone. Small veins of white calc-spar, crossing a bluish limestone passing from compact to granular, rather characterise the transition formation than the zechstein of the plains; but in both continents these small veins are also found in the limestone of the lofty calcareous mountains, which, by their position and their interposed beds of rock-salt and bituminous clay, belong, I believe, to zechstein. Besides, in all the formations above the red sandstone we observe that (probably by a galvanic action) the darkish-grey limestones lose their colouring principle in the vicinity of the planes of stratification; this decoloration takes place in rocks *in situ*. The accumulation of carbon exists only in the centre of the beds, and it would seem as if the stone had been exposed to the contact of light and the oxigen of the atmosphere.

Of all secondary formations the zechstein is that of which the various strata have been the most minutely studied; it is that which has most contributed to give rise, in the north of Germany, the classic ground of geognosy, to the first precise ideas on the relative age of formations, and the regularity with which they succeed each other. The bituminous and copper slates of the zechstein being a very important object of mining, it was necessary to pierce through five formations, muschelkalk, fibrous and clay gypsum, variegated or oolitic sandstone, lamellar and saliferous gypsum, and zechstein, to reach the argentiferous bed placed between the zechstein and the red sandstone. We may assert that the labours of miners in the bituminous slates of Mansfeld in Germany, and on the coal-measures in England, have assisted much the progress of the *geognosy of position*, of which Stenon has the honour of having first indicated the true principles.

Zechstein, or alpine limestone, the most ancient of the secondary formations, contains, as subordinate beds, slaty carburetted and bituminous clay, coal, rock-salt, gypsum, fetid limestone, compact or disintegrated (asche) magnesian limestone, limestone with gryphites, ferriferous limestone (eisenkalk), cellular limestone with crystalline grains (rauchwacke), sandstone, calamine, lead, hydrated iron and mercury. We shall add to these indications the substances that we sometimes found disseminated in zechstein, without forming continuous beds, such as sulphur, flint (hornstein), and rock crystal. In the whole of these masses three series, the bituminous or carburetted, muriatiferous, and metallic, are easily distinguished. The copper-slate containing petrified fish, fetid limestone, rock-salt, and gypsum, calamine and sulphuret of lead, are the most important types of these three series, and serve, in a certain degree, by their *geognostic concomitance*, to identify the formation which we are describing, where the relations of position are doubtful.

Schistose, carburetted, or bituminous clay or marl.

The accumulation of carbon which characterises the transition formations, particularly the latest, attains its maximum in the red sandstone; the carbon no longer appears as graphite or anthracite, but as bituminous coal. The formation of alpine limestone, so intimately connected with that of the red sandstone, or coal-sandstone, participates to a certain point in this abundance of hydrogenetted carbon; sometimes it is the whole mass of the rock (south of Bavaria, and Merlingen on the lake of Thun; in South America, mountains of New Andalusia) which is penetrated by bituminous parts; sometimes it is only interposed beds of clay and marl that contain bitumen. The most celebrated of these beds is the copper-slate (kupfer-schiefer) of Mansfeld, which in the new world contains fossil fish near Ceara (plains of Brazil), near Pasco (at 2000 toises high; Andes of Peru), near Mondragon (table-land of Potosi), and near Pongo de Lomasiacu (banks of the Amazon, province of Jaen). There is most frequently but one bed of copper-slate, and that bed is rather placed towards the lower limit of the zechstein. On account of this position it was long taken for an independent formation placed between the zechstein and the red sandstone. At other times (Conradswalde, Prausnitz, and Hasel, in Silesia) several beds occur which alternate with beds of zechstein, and are equally worth working. Argentiferous copper and lead are found only accidentally accumulated in this partial formation, and I saw in the two continents (Chiemsée and

Wallersée in the south of Bavaria; mines of Tehuilotepec in Mexico, mountain of Cuchivano near Cumanacoa) this copper-marl of Mansfeld represented by small beds of schistose, carburetted, darkish-brown clay, containing but little bitumen, and abounding in pyrites. This phenomenon appears to connect the zechstein of the plains with that of the high mountains, the superposition of which on the coal-sandstone is less evident. In the Andes of Montan (1600 toises high; the north of Peru) black clay from five to eighteen inches thick alternates with zechstein. Slaty and marly clay oscillates from zechstein or alpine limestone, on one side towards the red sandstone and transition limestone, and on the other towards Jura limestone. Argentiferous copper-slate is found again in the red sandstone, but with a great accumulation of carbon (Suhl and Goldlauter in Saxony). In the transition limestone (Schwatz in Tyrol) the clays become more micaceous, and pass to transition clay-slate containing (Glaris), like the slates of zechstein (Eisleben), and like those of the red sandstone (mine of Saint Jacques near Goldlauter), petrified fish. The marl in the Jura limestone is more calcareous, of a lighter colour, being whitish and bluish grey. Notwithstanding the analogies which the slaty and highly carburetted clays of the zechstein have with those of the coal-sandstone, it is only in the latter, which immediately cover the coal, that we find the impressions of the true ferns of the polypodiaceous group. Copper-slate exhibits only lycopodiacea, a family which Swartz has long separated from fern.

Coal. We have just stated, that a great accumulation of carbon particularly characterises the formation of red sandstone, but bitumen characterises the formation of alpine limestone; the latter however presents traces also of true coal, either in beds (between Nalzon and Pereilles in the Pyrenees; at Huanuco in the Andes of Peru, at 2000 and 2200 toises high) or as disseminated parts in copper-slate (Eisleben, Thalitter in Saxony). It is a remarkable fact, which has been long observed, that piciform coal (jet) appears most commonly on the impressions of the bodies of petrified fish; it replaces, in those organic impressions, sulphuret of iron, and (between Mörsfeld and Münsterappel, in the duchy of Deuxponts) native mercury and cinnabar. Beds of coal, mixed with marine shells and amber (Hering and Miesbach in Tyrol; Entrevernes on the lake of Annecy in Savoy) are not found in the zechstein; but only lignites that belong to much more recent formations. They are superposed to the zechstein in insulated basins, and, like all local formations, have their sandstones and clays.

Rock-salt and muriatiferous clay. The masses of rock-salt in alpine limestone or zechstein are subordinate, not to beds of lamellar gypsum, but to a particular formation of clay, which was long neglected by geognosts, and which I made known by the name of *salzthon* (muriatiferous clay). In both continents it characterises the deposites of rock-salt, as schistose clay (schieferthon) or *clay* with impressions of fern characterises the deposites of coal. This muriatiferous formation, in which gypsum may be said to be found only accidentally, was the principal object of my researches in the journeys which I undertook by order of the Prussian government, during the years 1792 and 1793, in the mines of rock-salt of Switzerland, the south of Germany, and Poland. I again found it with all its analogies in the Cordilleras of equinoxial America, and it cannot be doubted that the knowledge of its aspect is an object of the greatest interest to those who search for the discovery of deposites of salt in countries where it has hitherto been supposed wanting.

The colours of muriatiferous clay are generally (Hall, Ischel, Aussee) smoke-grey, and whitish and bluish-grey (Berchtolsgaden and Wieliczka); sometimes this clay is darkish-brown, reddish-brown (leberstein of the miners of Tyrol and Styria), and even brick-red. It is found in vast masses, or disseminated in small rhomboidal portions, either in rock-salt (Zipaquira in New Grenada), or in gypsum (Neustadt an der Aisch in Franconia, Reichenhall in Bavaria) which is subordinate to the alpine limestone. The colours of the muriatiferous clay are much more mixed and varied than those of the slate-clay that covers coal. The former produces a slight effervescence with acids; its colours are owing both to carbon and to oxid of iron. I saw them on the table-land of Bogota mixed with asphaltum, and staining the fingers black. It rapidly absorbs the oxigen of the atmosphere, either in our laboratories or in those great circular excavations (Sinkwerke, Wöhre) which are filled with fresh water for washing the saliferous rock. Its consistence is extremely variable, changing from soft to the hardness of copper-slate. Tenacious masses (schlief) often appear mixed with flint, and give sparks with steel; their fragments are then testaceous and curved (krummschalig abgesonderte Stücke). Being imbedded in a friable clay, these masses form a kind of porphyroid breccia. Muriatiferous clay contains neither scales of mica, nor impressions of fern, so common in the slate-clay of the coal mines; we, however, sometimes find pelagic shells in it (Hallstadt, Wieiczkla).

Rock-salt occurs in two ways; either disseminated in parts more or less visible in the salzthon, or forming thick beds alternating with beds of clay. This different disposition determines the *maximum* (Wieliczka) or the *minimum* (Ischel) of the richness of the mines; it decides whether the salt should be worked in great masses (as Pliny expresses it, *lapidicinorum modo cæditur sal nativum*), or by washing the rock by the introduction of fresh water into the subterraneous chambers. Even when the muriate of soda, of a smoke-grey colour, is disseminated in rounded grains, or in small plates, or in a manner invisible to the eye, it forms continued crusts around the *separate pieces* of salzthon. It fills all the crevices which divide the masses into polyhedral fragments; thence result argillaceous breccias (Haselgebirge) cemented by rock salt. Sometimes great masses of clay (Hall in Tyrol) are altogether destitute of muriate of soda; they are thought to have been washed by the action of the waters that circulate in the earth, and this curious phenomenon seems to favour the hypothesis originally adopted on the origin of salt-springs.

Granular gypsum, greyish-white, but rarely anhydrous (muriacite), occurs in beds more or less thick in *salzthon;* it abounds there more than in rock-salt, and its volume is far inferior to that of clay. Sometimes the gypsum is mixed with fetid limestone and crystals of magnesian carbonate of lime (ranten ou bitterspath). When the salt does not form true beds or crystalline continued masses, it is found like an *interwoven mass* in the clay (Stockwerk), that is, in small veins that *cross* each other, *swell*, and *extend* in every direction. These fibres are perpendicular to the wall and ceiling of the veins (Berchtolsgaden). At other times the salt is divided into very thin beds, parallel to each other, varied in colour, sinuous, and generally vertical (Hallstadt and Hallein), rarely inclined less than 30° (Aussee). Wherever granular gypsum is entirely wanting in the *salzthon* it is replaced by scattered crystals of selenite. The whole of this saliferous formation contains occasionally disseminated pyrites, brown blende, and galena. At Zipaquira in South America (mine of Rute), pyrites and ferriferous carbonate of lime form particular concretions in flattened spheroids, from 18 to 20 inches diameter, containing crystallised spathic iron. I did not observe this singular phenomenon in the mines of rock-salt of Germany, Poland, and Spain, which I visited; but the frequency of pyrites in muriatiferous clay throws some light on the small sulphuretted hydrogen that so often exhales from salt-springs. Galena appears only in small masses in the saliferous deposite of Hall in Tyrol; but is developed in great masses in the mountains of rock-salt (reddish-white and darkish-grey) across which, at a distance of two leagues, the Rio Guallaga and the Rio Pilluana have cut their way (Peruvian province of Chachapoyas, on the eastern declivity of the Andes).

The deposites of salt in the two continents are generally found uncovered, like the formations of euphotide and serpentine. Sometimes they support small beds of gypsum and fetid limestone, which belong to them exclusively. It is consequently not easy to decide on the relative age of muriatiferous deposites. The principal formation (Hauptsalzniederlage) appears to me to belong evidently to the zechstein, or alpine limestone; but this does not exclude the probability that other partial formations are interposed in the transition and perhaps even in the tertiary formations. Coal, oolite, and lignite, are also developed at very different periods from each other, and yet the principal positions of these three substances are the red sandstone, Jura limestone, and plastic clay. In order to treat this subject in the most general manner, I shall indicate successively, according to the actual state of our knowledge, the various formations of rock-salt in the transition limestone, zechstein, and variegated sandstone with clay.

The anhydrous gypsum of Bex, which contains disseminated rock-salt and small subordinate beds of grauwacke, belongs, according to the observations of MM. de Buch and Charpentier, to transition limestone, but probably to the last beds of the intermediary formation. The saliferous gypsum of Colancolan (at the east of Ayavaca, Andes of Peru) appears mixed, like the transition limestone of Drammen (Norway), with tremolite asbestoid; the small deposites of S. Maurice (Arbonne in Savoy), and, according to M. Cordier, the mountain of salt, at Cardona in Spain. Anhydrous gypsum particularly characterises the saliferous deposites of the transition formation. In the south of Germany, on the banks of the Necker (Sulz above Heilbronn; Friedrichshall, between Kochendorf and Jaxtfeld; Wimpfen, above Heilbronn), rock-salt has been discovered in zechstein, by means of soundings of 245 and 760 feet deep. The admirable labours of MM. Glenk and Langsdorf leave no doubt on this subject. At Sulz, they have successively pierced through muschelkalk, the formation of clay and variegated sandstone, porous zechstein of small thickness, and red sandstone reposing on the granite of Bergstrasse and Schwarzwald. At Friedrichshall and Wimpfen, according to the judicious observations of M. de Schmitz, the upper beds of the zechstein are entirely wanting, and in what exists, which is bluish-grey, and which has on that account been often confounded with transition limestone, alternating beds have been found of rock-salt, saliferous clay, and white and greyish gypsum. In the Grand duchy of Baden, the saliferous deposite appears covered (Heinsheim near Wimpfen, on the Necker; Stein, Mühlbach and Beyerthal, in the valley of the Rhine; Kandern, in Schwarzwald) by the same rocks which occur at the salt-work of Sulz.

I may also mention as a very evident proof of the position of the great formation of rock-salt in the zechstein or alpine limestone, the northern part of the table-land of Santa Fé de Bogota, where the mine of Zipaquira is found (Rute, Chilco, and Guasal) at 1380 toises above the level of the sea. This saliferous deposite, more than 130 toises thick, is covered by great masses of granular gypsum which is seen interposed, in several places very near to the mine, in zechstein supported by the red sandstone, or coal-sandstone. There is only a distance of seven leagues from the coal mine of Canoas, and the mine of rock-salt of Zipaquira. Other deposites of coal (Suba, Cerro de Tunjos) are still nearer, and we see red sandstone, which is very quartzose, appear immediately from beneath the saliferous clay of Zipaquira.

Since I first visited Salzbourg, the Tyrol, and Styria, no doubt has remained on my mind of the close connection of rock-salt with zechstein.

Many celebrated geognosts (MM. de Buch and Buckland) are of the same opinion; but it must be admitted, that wherever the age of the limestone is not sufficiently characterised by the presence of coal-sandstone, and wherever the covering of the saliferous deposite by beds of a known age is not evident, observations do not afford complete conviction. In the mine of Hall, near Inspruck, we see (gallery of Mitterberg) the deposite of rock-salt immediately covered by the limestone formation that constitutes the northern chain of the Alps of the Tyrol. This limestone passes from greyish-white to greyish-blue; the dark-coloured parts are often fetid. It is generally compact, sometimes rather fine-grained granular, and traversed by veins of white calc-spar. These veins are considered by some geognosts, and perhaps too particularly, as characterising the transition limestone. The rock no where alternates either with intermediary clay-slate or grauwacke; it forms (Wallersée) contorted beds, like the limestone of the lake of Lucerne. M. de Buch has frequently found in it petrifactions of very small turbinites. This is the only place in Europe where I saw a considerable limestone formation immediately covering rock-salt. I believe it to be zechstein, from its position and structure; I have seen it sometimes pass (Schlossberg, near Seefeld; Scharnitz) to compact limestone, with a dull fracture, even, or conchoidal, with very flat cavities, similar to the lithographic limestone of Jura (lias). The petrified fish that are found between Seefeld and Schönitz, in a bituminous marl, remove the limestone of Hall still farther from transition limestone; in order, however, to characterise it indubitably as zechstein, we must find it reposing on the red sandstone (todtliegende), which, according to the observations of MM. Uttinger and Keferstein, appears to be superposed on the intermediary rocks between Ratenberg and Hering, as well as near the ancient mines of Schwatz. M. de Buch and myself observed, at Hallstadt (Törringer Berg) and at Ischel, alpine limestone analogous to that of Hall, but with lighter colours, often reddish, and more abundant in petrifactions, superposed on gypsum which covers deposites of rock-salt. This superposition is less evident at Hallein (mine of Durrenberg) and at Berchtesgaden; the gypsum which covers the saliferous clay is hid beneath a calcareous puddingstone (nagelfluhe) of tertiary formation. The deposites of Hallein and Berchtesgaden appeared to me like that of Wieliczka in Poland, not interstratified with zechstein, but superposed on that formation. I believe them to be posterior to the great formation of coal; but the red sandstone is wanting in their vicinity, and the limestone of the country of Salzbourg is immediately superposed (valley of Ramsau) on grauwacke. M. Buckland considers the limestones that cover the saliferous clay, at Hallstadt, and even at Bex, as belonging to the lias, which is the lower bed of Jura limestone.

After the rock-salt of the anhydrous gypsum

of transition, and that of zechstein, comes, according to the age of formation, the salt of the variegated sandstone, or, to speak more correctly, of the formation of clay and variegated sandstone. This arenaceous formation, called by the English geognosts new red sandstone and red marl, contains the deposites of salt (Norwich) of England; it contains salt also in Germany, near Tiede (between Wolfenbüttel, and Brunswick), where MM. Haussmann and Schulze found small masses of disseminated salt in the red clay of the variegated sandstone of the oolite; at Sulz (kingdom of Wurtemberg), where, before they reached the salt-springs in zechstein, they found, immediately below the muschelkalk, at the depth of 460 feet, nodules of salt in the red marl. This clay, in a thickness of 210 feet, covers the variegated sandstone to which it belongs. As rock-salt alternates, very near Sulz (at Friedrichshall, and Wimpfen), with marl and gypsum interstratified with zechstein, we cannot doubt the geognostic affinity that exists between the two formations of zechstein, and variegated sandstone. Marl and saliferous clay with granular gypsum is found placed sometimes between the zechstein and sandstone, sometimes in other of those formations. To this formation of clay and variegated sandstone belongs also the rock-salt of Pampeluna in Spain, examined by M. Dufour, and the rich deposite discovered in 1819, in Lorraine, near Vic. This variegated clay formation of Vic contains small beds of muschelkalk, and is covered in its turn by Jura limestone. The influence which a more accurate knowledge of the position of rocks has had in these latter times on the discoveries of salt in Swabia, France, and Switzerland, (Eglisau, canton of Zurich), is a phenomenon well worthy of observation.

I doubt if we have hitherto very certain proofs of the presence of rock-salt in muschelkalk; for we must not, as we shall soon see, deduce this position only from the existence of salt-springs. Muschelkalk, in its lower beds, alternates with the formation of *clay and variegated sandstone;* as it also contains sometimes (Sulzbourg, near Naumbourg) marl with fibrous gypsum, it would not be surprising if we discovered in it some saliferous deposites. Traces of these deposites have been observed, near Kandern, in Jura limestone.

Do beds of salt exist in the tertiary formations above the chalk? Several geognostic phenomena may lead to this supposition, and we may almost wonder that the last irruptions of the ocean on the continents have not produced, if not beds of rock-salt, at least of saliferous clay. In the actual state, however, of our knowledge, this problem is not sufficiently determined. M. Steffens considers the gypsum with boracite of Lunebourg and Seegeberg, (Holstein) as above the chalk. The latter of these gypsums contains small masses of disseminated rock-salt; the former gives rise to very rich and abundant salt-springs. Other geognosts believe the gypsum formation with boracites to be much more ancient than the gypsum with bones of the tertiary rocks, and almost identical with the gypsum of the zechstein and variegated sandstone. The immense saliferous deposites of Wieliczka and Bochnia, which extend from Galicia as far as Bukowine and Moldavia, appear to repose immediately on coal-sandstone, containing at the same time, which is a remarkable fact, anhydrous gypsum, tellinæ, univalve, polythalamous shells, fruits in a carbonated state, leaves, and lignite; these deposites are covered only by sand and micaceous sandstone. M. Beudant, in his important work on Hungary, seems to lean towards the opinion that these sands and sandstones are analogous to the molasse of Argovia; and that all the saliferous formations of Galicia may be contemporary with the plastic clay (sandstone with lignites) of the tertiary rocks, placed between the chalk, and the Paris limestone, (limestone with cerithia). This bituminous wood of Wieliczka, exhaling the smell of truffles, merits, no doubt, great attention; and if we admit that it is only mixed accidentally with rock-salt, and that it comes from superposed sandy beds, we must still conclude that rock-salt and these sands are nearly connected in their origin. But is the presence of lignite a very convincing proof of the late origin of a bed? This I doubt: we know that lignite and the impressions of dicotyledon leaves are found far below the chalk, and in the lower beds of the Jura limestones (limestone with gryphæa arcuata; Vay, Issigny near

Caen), in quadersandstein, and in small carbonated and marl beds (lettenkohle) of muschelkalk, and in the variegated sandstone of Germany, to which also belongs the argentiferous slates of Frankenberg (Hesse). We must carefully distinguish the siliceous and petrified wood from lignite or bituminous wood (braunkohle); and if we rarely find the latter in the clay of the variegated sandstone, we find it still less in zechstein, of which the copper-slate alone contains petrified fruits. In Tuscany, according to M. Brongniart, the salt-springs of Volterra spring from the marly beds that alternate with granular gypsum (alabaster), and which are immediately covered by a tertiary formation. Although it seems almost impossible to decide on the age of *formations which are not covered*, several relations of position which I had occasion to observe in the new continent induce me to think it probable that deposites of salt exist in tertiary formations. I shall not cite the mountains of rock-salt in the vast plains at the north-east of New Mexico, which Mr. Jefferson first made known, and which appear to be connected with the coal-sandstone; but other very doubtful deposites, such as the saliferous clay, superposed on the trachytic conglomerates of the Villa d'Ibarra (table-land of Quito, at the height of 1190 toises), the marly masses of salt worked at the surface (deserts of Lower Peru and Chili), in the steppes of Buenos-Ayres, and in the arid plains of Africa, Persia, and Transoxane. I saw near Huaura (between Lima and Santa, on the coast of the South Sea), rocks of trachytic porphyry pierce through beds of the purest rock-salt. The muriatiferous clay of Araya (gulf of Cariaco), mixed with lenticular gypsum, appears to be placed between the alpine limestone of Cumanacoa, and the tertiary limestone of Barigon and Cumana. The salt is accompanied on all these points with petroleum and solid asphaltum.

In comparing the deposites of rock-salt of England (at 30 toises), of Wieliczka (160 toises), of Bex (220 toises), of Berchtolsgaden (330 toises), of Aussee (450 toises), of Ischel (496 toises), of Hallein (620 toises), of Hallstadt (660 toises), of Arbonne in Savoy (750 toises?), and of Hall in the Tyrol (880 toises), M. de Buch has judiciously observed, that the riches of these deposites diminish in Europe with the height above the level of the ocean. In the Cordilleras of New Grenada, at Zipaquira, immense beds of rock-salt occur, not interrupted by clay, at the height of 1400 toises. The mine of Huaura only, on the coast of Peru, appeared to me to be richer; I there saw salt worked in the same manner as a quarry of marble.

In Thuringia, one of the countries where the succession and the relative age of rocks was first observed, it was long believed that salt-springs are more frequent in the granular gypsum of the zechstein than in the fibrous and clay gypsum of the variegated sandstone; and the former exclusively was regarded as saliferous. The natural caverns in the inferior gypsum (salzgyps and schlotten-gyps) have always been considered as cavities formerly filled with rock-salt. In hazarding these hypotheses, founded on too few observations, it was forgotten that the deposites of salt are much less characterised by granular gypsum than by a clay (salzthon) very analogous to the clay of the upper or fibrous gypsum. Salt-springs occur together in groups, or succeed each other in sinuous and diverging bands. The direction of these subterranean rivers appears to be independent of the inequalities of the surface of the soil. Such is the circulation of the waters in the interior of the globe, that the saltest springs may often be the farthest removed from the spot where they dissolve rock-salt. A great degree of saltness does not prove the proximity of that cause, any more than the violence of earthquakes proves the proximity of volcanic fire. The springs are sometimes plunged into the lower beds, and sometimes by hydrostatic pressure rise towards the upper beds. It is not their position solely that can throw light on the source of saliferous deposites. We find salt-springs in Germany in grauwacke-slate of the transition formation (Werdohl in Westphalia), in the porphyry of the red sandstone (Creuznach), in the red sandstone itself (Neusalzbrunnen near Waldenburg), in the gypsum of zechstein (Friedrichshall near Heilbronn, Wimpfen on the Necker, Dürrenberg? in Thuringe), in the formation of clay and variegated sandstone (Dax in France;

Schönebeck, Stasfurth, Salz der Helden in Germany), and in the muschelkalk (Halle? in Saxony; Süldorf, Harzburg). We may add to this enumeration the Jura limestone (Butz in Frickthal), and perhaps the molasse (tertiary sandstone with lignite) of Switzerland (Eglisau, soundings of M. Glenck). We must not, in the search for rock-salt, mistake for true deposites, small masses which highly concentrated salt-springs may have accidentally left by evaporation on the clefts of the rocks.

Gypsum and fetid limestone. Some formations of gypsum posterior to transition gypsum (§ 20.) appear in all limestone formations above the red sandstone, in the zechstein, in the red sandstone itself, in muschelkalk (very rarely), in the Jura limestone, and in tertiary formations. The gypsum (unterer gyps, schlolttengyps of Werner) which belongs to the zechstein occurs less in extended beds than in irregular masses; it is often superposed (Thuringe) on zechstein, and covered by the variegated sandstone. It is compact or granular, and alternates with fetid limestone (stinkstein), while the gypsum of the variegated sandstone (oberer gyps, thongyps of Werner), is rather fibrous and mixed with clay. These characters of structure and mixture are, however, not general. We stated above, that in the saliferous gypsum of the zechstein, clay (salzthon) takes an extraordinay developement. On the other hand, the fibrous and clay gypsum of the variegated sandstone sometimes exhibits also granular masses (alabaster of Reinbeck, in Saxony), breccia of fetid limestone, and considerable cavities (gypsschlotten): three phenomena which more generally characterise the gypsum of the zechstein.

All these phenomena prove the close connection that exists between the two great saliferous formations, alpine limestone and variegated sandstone with clay. I saw frequent examples, in the equinoxial zone of the new continent, of beds of gypsum interstratified with or superposed on zechstein, in the Llanos of Venezuela (Ortiz, Mesa de Paja, Cachipo); in the province of Quito (table-land of Cuença, near Money, and between Chulcay and Nabon); in the table-land of Bogota (Tunjuellos, Checua, and at 1600 toises above the level of the sea, at Cucunuva); in the plains of the Amazon (Quebrada turbia near Tomependa); at Mexico, between Chilpansingo and Cuernavaca (near Sochipala); and in the metalliferous mountains of Tasco and Tehuilotepec.

The beds of fetid limestone are either subordinate to gypsum and muriatiferous clay containing zechstein, or they appear to be the result of an accidental accumulation of bitumen in the rock of zechstein itself. This accumulation gives rise to springs of mineral pitch, and perhaps also to those flames of hydrogen that issue from the alpine limestone in Europe, in the Apennines (Pietra Mala, Barigazzo), and in America, in the mountains of Cumanacoa (Cuchivano, lat. 10° 6′). Fetid limestone occurs also, but much more rarely, in the variegated sandstone and muschelkalk (beds of belemnites of Gœttingue?). The asche and the *rauhkalk* of the mines of Thuringia are only pulverulent or crystalline and porous varieties of fetid limestone belonging to the zechstein. As fetid limestone is in Europe constantly destitute of petrifactions, I shall here mention that in the plains of New Grenada (valley of the Rio Magdalena, between Morales and the mouth of the Caño Morocoyo), M. Bonpland found terebratulites and pectinites in a variety of this rock which was dark-grey, a little brilliant at the exterior, strongly bituminous, and traversed by veins of white calcareous spar.

Magnesian limestone. We must distinguish in geognosy, between beds interstratified with zechstein (gypsum, rock-salt, sulphate of lead), the chemical composition of which differs entirely from that of the principal rock, and the partial modifications of the rock itself. The modifications which affect the structure (the grain more or less crystalline, the oolitic form, or the porosity) and the mixture (magnesian limestone, ferriferous limestone) are less important than might at first be supposed. We find analogies in formations of a very different age; they characterise certain formations in cantons of small extent, but when we compare very remote regions, we see that they do not characterise them as much as interstratified beds that are chemically heterogeneous. In England the great mass of magnesian limestone, red-land-limestone of M. Smith, often contains many madrepores (Mendip hills near Bristol) and connected with limestone-breccia, or cellular beds (Yorkshire) similar to rauchwacke, is no doubt the parallel of the zechstein; it is placed between the formations of coal and rock-salt: in England, however, as well as in some parts of the continent, according to the researches of MM. Buckland, Brongniart, Beudant, Conybeare, Greenough, and Phillips, the mixture of magnesia and carbonate of lime, which Arduin observed in the Vicentine in the year 1760, is also found in the variegated sandstone with clay (red marl), in the oolitic limestone of Jura, in the chalk, and in the calcaire grossier (Parisian limestone) of tertiary formation. Perhaps even in Hungary and in a part of Germany, the magnesian limestone belongs rather to the variegated sandstone and to the oolitic formations of Jura than to the zechstein. These rocks are in general straw-colour (of Sunderland and Nottingham) or reddish-white, sometimes compact, sometimes rather granular, pearly and shining in the fracture; sometimes they are cellular, and traversed by veins of calc-spar. They effervesce slowly with acids, and, like the true dolerite of primitive formations, they often form but thin beds in a limestone which is not magnesian. If in magnesian limestone and red marl with rock-salt (two formations placed between the coal and the oolitic deposites), we recognise in England the zechstein and the variegated sandstone of the continent, we must not forget that in Germany and Hungary, zechstein is connected with the red sandstone or the coal-sandstone; while in England, the coal deposite is generally found in a position non-conformable with those of the magnesian limestone, and almost belonging to the transition rocks. The *three great deposites* of *coal*, *salt*, and *oolites*, which serve in a degree as marks to the geognost while he is exploring an unknown country, are every where placed in the same manner, but the mutual links of formations, and the degree of their developement, vary according to the localities. In England, when by the suppression of the *new red conglomerate* (todtes liegende), magnesian limestone (zechstein) reposes immediately on the coal deposite (Durham, Northumberland), the coal is regarded as of an inferior quality.

Ferriferous limestone, rauchwacke, and limestone with gryphites. The ferriferous limestone (eisenkalk, zuchtwand) is a brownish or dull-yellow rock, sometimes compact, sometimes granular or cavernous, penetrated with spathic iron, forming beds in the upper layer of zechstein (Cammsdorf, Schmalkalden, Henneberg). It is sometimes traversed by copper-slate, and takes such a developement that it replaces all the lower beds of zechstein. When it becomes darkish-grey, charged with bitumen and cavernous, it bears the name of *rauchwacke* in Germany. The cavities of rauchwacke are angular, long, and narrow, and covered with crystals of carbonate of lime. This small partial formation, which M. Karsten, in his *Classification of Rocks*, confounded with the cavernous and porous part of the Jura limestone, is sometimes magnesiferous, imperfectly oolitic (Cresfeld), and mixed with granular quartz. Fetid stone, ferriferous limestone, and rauchwacke are closely connected together. To rauchwacke also, for the most part, that vast collection of gryphite belongs (*G. aculeatus*) which is called *limestone with the gryphæa aculeata* (gryphitenkalk), which characterises the zechstein, and which, as we shall afterwards see, forms a bed more ancient than the *limestone* with the *gryphæa arcuata*, one of the lower strata of the Jura limestone.

Sandstone. Wherever zechstein or alpine limestone is developed alone in great masses, and is consequently not interstratified with red sandstone, beds of sandstone are very rare. I observed some, however, in the mountains of Cumana (Impossible, Tumiriquiri). This sandstone interstratified with zechstein is extremely quartzose, destitute of petrifactions, and alternates with darkish-brown clay. M. de Buch observed in Switzerland a phenomenon altogether analogous, in the alpine limestone of Molesson, and in that of Jaunthal near Fribourg. In the Cordilleras of Peru near Huancavelica, more than 2000 toises above the level of the ocean (mine of Santa Barbara), an immense bed of sandstone as quartzose as that of Fontainebleau, and containing a deposite of mercury, forms a bed in alpine limestone. Even the zechstein of Thuringia sometimes contains small beds of sandstone extremely quartzose, that traverse the copper-slate. An arenaceous marl (weissliegende) occurs on the junction of the zechstein with the red sandstone. It varies much in its composition, and reminds us of the beds of sandstone of Tumiriquiri in South America. The weissliegende of Thuringia is generally calcariferous, and contains sandstone and siliceous conglomerates. M. Freiesleben found in it (Helbra) globular concretions similar to those which I collected in the saliferous clay of the zechstein of Zipaquira. We shall again observe, that the alpine limestone of the Pyrenees is not only mixed with sand and mica, but also contains beds of argillaceous sandstone.

Sulphuret of lead, hydrate of iron, calamine, mercury. These four small metallic formations characterise the zechstein in the two hemispheres. Argentiferous galena appears already in small masses in the copper-slate of Thuringia, but in Silesia and Poland it forms (Tarnowitz, Bobrownik, Sacrau, Olkusz, Slawkow) very extensive beds in zechstein, consequently above the rich deposite of coal of Ratibor and Beuthen. In the same country beds of hydrate of iron (Radzionkau) and calamine (Pickary), parallel to each other, are of a much more recent origin than the bed of argentiferous sulphuret of iron of Tarnowitz.

We already find in the granular limestone destitute of shells, which covers the latter bed, small masses of brown iron and concretioned oxide of zinc disseminated in elongated cavities. Near Ilefeld in Hartz, the whole of the zechstein is impregnated with this latter substance. With respect to the beds of galena and calamine of Sauerland, Brilon, Aix-la-Chapelle, and Limbourg, they seem, according to the judicious observations of MM. de Raumer and Nœggerath, notwithstanding their apparent analogy with the formations of Upper Silesia, to belong to the most recent transition rocks. It might be said, there exists in the two continents a *geognostic affinity* (or position) very remarkable between limestone rocks and sulphuret of lead which is more or less argentiferous; we see the latter in Europe in intermediary limestone (veins of Schwatz in the Tyrol, and mountain limestone of Northumberland, York, and Derbyshire), and in the alpine limestone (beds of Upper Silesia and Poland; magnesian limestone of Durham). On the table-land of New Spain, the lead ores of the district of Zimapan (Real del Cardonal, Lomo del Toro), as well as those of Liñares and New Saint Ander, belong also to the limestone mixed with fetid stone, and which immediately succeed the coal formation.

Calamine occurs in the magnesian limestone of England (Mendip hills), as well as in the zechstein of Upper Silesia. With respect to the argillaceous beds of hydrate of iron, they exhibit a peculiar character in the alpine limestone of the Andes of Peru;

they contain abundance of native filiform silver, and muriate of silver. This mixture of oxides of iron and of silver which M. Klaproth and myself made known, bears the name of *pacos ;* it is found in the equinoxial part of both Americas, occupying the upper part of the veins, and it exhibits in this position a very remarkable analogy with the earthy ochraceous (but not argentiferous) masses which the miners of Europe call vulgarly by the name of the *iron cap* (chapeau de fer) of the veins (eiserne Hut). The richest example that I am acquainted with of a *bed of pacos* in the alpine limestone, is the deposite of the mountain of Yauricocha (Cerro de Bombon, Peruvian Cordillera of Pasco), situated at more than 1800 toises of absolute height. Although the workings of this place, of oxide of iron abounding in silver, have in general reached hitherto only to the depth of fifteen to twenty toises, they have furnished, in the last twenty years of the eighteenth century, more than five millions of marks of silver. An experienced geognost will consider this remarkable position only as a particular developement of the beds of hydrate of iron that occur in the zechstein in Upper Silesia, and sometimes pass (Pilatus and Wallensee in Switzerland) to lenticular iron.

The simultaneous occurrence of mercury in coal-sandstone and alpine limestone adds to the relations which we have indicated between those two formations. In Carniole (Idria), ores of mercury are found, according to MM. Héron de Villefosse and Bonnard, in marl-slate similar to the copper-marl of Mansfeld. At Peru, near Huancavelica, cinnabar is partly disseminated in a very quartzose sandstone that forms a bed in the alpine limestone (Pertinencias del Brocal, Comedio, and Cochapata, mine of Santa Barbara); a part fills the veins (mountain of Sillacasa) which unite into bunches and traverse the alpine limestone.

After having mentioned this great variety of true beds contained in the formation of which we endeavour to describe the relations of position, structure, and composition, it remains for me to indicate the substances which are found simply disseminated in it. I shall confine myself to naming flint, rock-crystal, and sulphur.

Common flint (hornstein), very rare in the zechstein of plains (Thuringia), characterises that formation in the alpine region of the Pyrenees, Switzerland, (Mont Bovon, la Rossinière) Salzbourg, and Styria (above Hallstadt, Potschenberg, Goisern); it often passes to jasper and gun flint (feuerstein). The flint of the alpine limestone is found in Europe, only in nodules often disposed on the same line; but, in the Cordilleras of Peru, in the rich silver mines of Chota (near Micuipampa, south lat. 6° 43′ 38″), flint forms a bed of immense thickness. The mountain of Gualgayoc, which rises like a fortified castle on a table-land 1800 toises high, is entirely composed of this substance. The summit of the mountain is terminated by an innumerable quantity of small pointed rocks, each having large openings, which the people call windows (ventanillas). The flint (*panizo*) of Gualgayoc is a scaly hornstone that is greyish-white, of a fracture dull and often even, and intimately mixed with sulphuret of iron. It passes sometimes to quartz, sometimes to gun flint. In the first case it is cellular, with irregular cavities, and covered with crystals of quartz. Great masses of this *panizo*, in which veins of grey and red silver and veins of magnetic iron form bunches of extraordinary richness, resembling the Parisian siliceous limestone of tertiary formation; but we see clearly, in several of these mines (Choropampa, at the east of Purgatorio, near the ravine of Chiquera), that this metalliferous hornstone is a bed of irregular form, interstratified with zechstein or alpine limesone. It contains imbedded great masses of limestone, and sometimes alternates (Socabon of Espinachi) with that very dark brown and slaty clay, which is found in the alpine limestone of Montan, and which renders the veins entirely sterile. The hornstone is destitute of shells, which abound in the principal rock, and which sometimes even fill the veins. An enormous mass of siliceous matter which appears as if melted in the midst of secondary limestone, with curved beds, and containing ammonites from eight to ten inches of diameter, is no doubt, a very remarkable geognostic phenomenon. Do nodules of hornstone (silex corné) exist (vicinity of Florence) in transition limestone? Of what age are the jaspers and disseminated calcedonies, in the Monti Madoni of Sicily?

The alpine limestone of Cumanacoa (South America) contains, like that of Grosörner (Thuringia,) disseminated rock-crystals. Those crystals are not found in cavities, but imbedded in the rock, like feldspar in porphyry, and like rock-crystal or boracite in gypsum of late origin.

Native sulphur, which we have already seen in the granular quartz of primitive formation, and in transition gypsum, (Sublin near Bex) re-appears in alpine limestone (Pyrenees, near Orthès, and near the forge of Bielsa; Sicily, Val de Noto and Mazzara), and in the foliated gypsum (New Spain, Pateje near Tecosautla) that belongs to this latter formation. The greater part, however, of the sulphur that abounds in the equinoxial regions of America, is found in trachytic porphyries, and in the clay of pyrogenic formation.

The operations of Bouguer and La Condamine having been carried on in a portion of the Andes where trachytic formations abound, among many erroneous ideas spread through Europe, on the structure of the Cordilleras, was that of the absence of shells and limestone formations in the equinoxial region. Even towards the end of the eighteenth century, the Academy of Sciences of Paris requested M. de La Peyrouse (Voyage, t. i. p. 169. "to examine if it was true that near to the line, or rather, as we approach towards it, that the calcareous mountains lowered, so as at last not to be above the level of the sea." In more recent works (Greenough, *Crit. Examination of Geology,*

p. 288.) doubts are mentioned of the existence of ammonites and belemnites in South America. In describing the superposition of rocks in different parts of the new continent, I indicated at what a prodigious height the shelly beds of zechstein rise in the Cordilleras of Peru and New Grenada. We must not suppose that the great revolutions which buried the pelagic animals were confined to particular climates.

In regions the most remote from each other, we find, in the formation of zechstein, or alpine limestone, gryphites (*G. aculeata*), entrochites (forming in many parts of Germany, according to the curious observations of M. de Buch, a distinct bed on the junction of alpine limestone and coal-sandstone), terebratulites (*T. alatus*, *T. lacunosus*, *T. trigonellus*); pentacrinites of great length; a trilobite in copper-schist, which, as to its genus, has perhaps not been sufficiently examined (*T. bituminosus*); ammonites (more rare than in muschelkalk and the marl of Jura limestone); some orthoceratites; fish which had attracted the attention of the ancients (Aristotle, *Mirab. auscultat.*, *ed Beckmanniana*, c. 75.; Livius, lib. 42. c. 1.); bones of the monitor, perhaps even (Tocayma and Cumanacoa, in South America) of crocodiles; impressions of lycopodiaceæ and bambousaceæ; no real fern, but, which is very remarkable (bituminous marl of Mansfeld), leaves of dicotyledon plants analogous to willow-leaves. It is observed that the shells of the alpine limestone (*Ammonites ammonius, A. amaltheus, A. hircinus, Nautilites ovatus, Pectinites textorius, Pectinites salinarius, Gryphites gigas, G. aculeatus, G. arcuatus, Mytulites rostratus*) are not so much disseminated in the mass of the rock as is the case with the two formations of muschelkalk and Jura limestone, as accumulated in certain parts, and often at great heights. Alpine limestones appear on a very considerable extent of country, sometimes entirely destitute of organic remains.

In the preceding pages we have indicated the formations of equinoxial America that belong to the zechstein. There are, in the littoral chain of the Caracas, the limestones of Punta Delgrada, Cumanacoa, and Cocollar, containing, not grauwacke, but quartzose sandstone and carburetted marl; in New Grenada, the limestone of Tocayma and the table-land of Bogota, supporting the rock-salt of Zipaquira; in the Andes of Quito and Peru, limestone of the province of Jaen de Bracomoros, Montan, and Micuipampa, placed on coal-sandstone, and containing enormous masses of flint; in New Spain, the limestone of Peregrino, Sopilote, and Tasco, between Mexico and Acapulco. Several of these calcareous masses of enormous thickness, and supporting gypsum and sandstone, are superposed not on coal-sandstone, but on transition porphyries very metalliferous and connected, at least in appearance, in some places, with a decidedly trachytic formation. It is observed in both continents, that, where alpine limestone has assumed a great developement, coal-sandstone is almost entirely wanting, and *vice versa*. This opposition in the developement of two neighbouring formations struck me, especially at Guanaxuato (central table-land of Mexico), and at Cuença (central table-land of Quito), where the coal-sandstone abounds; it arrested my attention in the Cordilleras of Montan (Peru), and at Tasco (New Spain), where alpine limestone abounds. We here repeat that when the coal-sandstone is not visible, or is not developed, the limits between alpine and transition limestone are very difficult to trace. In excluding from the secondary formation all the greyish-blue limestones traversed by veins of white calc-spar, and by beds of clay and marl, the formations of Cumanacoa, Tasco, and Montan (Venezuela, Peru, and Mexico), like those of the most northern Alps of the Tyrol and Salzbourg, become transition formations. I am inclined to think that the formations we have just named, like those of the Mole, Haacken, and Pilatus, are the most ancient beds of zechstein which are connected with the transition limestone of la Dent de Midi, Oldenhorn, and Orteler. Many rocks succeed each other by a progressive developement, and it appears quite natural that the last beds of a more ancient formation should exhibit a great analogy of structure with the first beds of the superposed formation.

It was recently proposed to place among the beds interstratified with zechstein or alpine limestone, two other substances, greenstone and dolerite, which we already know to be subordinate to coal-sandstone in many parts of Europe; even syenites, porphyries, and *secondary granites* have been indicated as superposed on alpine and Jura limestones. They are rocks of that part of the south-east of the Tyrol (valleys of Lavis and Fassa; Recoaro) on which Count Marzari Pencati has published very curious observations. The position of these substances being still a contested point of geology, I ought here to confine myself to mentioning the state of the problem, and of a question so worthy of the attention of geognosts.

M. de Buch already remarked in 1798, that between Pergine and Trento (Lago di Colombo, Monte-Corno), transition porphyry (or rather that of the red sandstone) alternates with alpine limestone of secondary formation. This limestone abounds in ammonites and terebratulites. The alternation is evident, and the porphyries so common every where else in coal-sandstone extend as far as the alpine limestone, in the same manner as on the eastern ridge of the Andes of Peru (Chamaya) I saw the compact quartz-rock, representing coal-sandstone, reach also to this very formation. It is a *penetrating* of the lower formation into a superposed formation, a phenomenon which may surprise us so much less, as in Silesia, Hungary, and in several parts of equinoxial America, the red sandstone, or the coal-sandstone, is closely connected with zechstein. The porphyries of the south of the Tyrol rise (mountain of Forna) to the height of 1500 toises. (Buch, *Geogn. Beob.*,

t. i. p. 303. 309. 315, 316.) M. de Marzari, whose researches began in 1806, thinks he saw in the vicinity of Recoaro, succeeding each other from below upwards, mica-slate and dolerite (filling at the same time the veins that traverse the mica-slate, and containing pyroxene and titaneous iron); red sandstone with coal and bituminous marl; zechstein, of which the lower beds are a limestone with gryphites, a formation of syenitic porphyry with interposed amygdaloid. M. de Marzari indicates, in the valley of Lavis (Avisio), always from below upwards, grauwacke, porphyry, red sandstone, alpine and Jura limestone, granite, and black pyroxenic masses destitute of olivine. According to the interesting memoir published by M. Breislak, the secondary granite placed on alpine limestone is altogether similar to the finest granite of Egypt; it contains (Canzacoli delle coste, Pedrazzo), *large masses of quartz with tourmaline;* its contact renders the limestone that supports it granular (at the depth of several toises); it passes sometimes to a *pyroxenic rock*, sometimes to a porphyry with a black feldspathic base, sometimes to *serpentine*. (Marzari, *Cenni geologici*, 1819, p. 45.; Id. *Nuevo osservatore Veneziano*, 1820, No. 113. et 127.; Breislak, *Sulla giacitura delle rocce porfiritiche e granitose del Tirolo*, 1821, p. 22. 25. 52.; Marzari, *Lettera al signor Cordier*, 1822, p. 3.; Maraschini, *Obs. géogn. sur le Vicentin*, 1822, p. 17.) Between la Piave and the Adige, an agathiferous mandelstein, which reminds us of those of the red sandstone, surmounts alpine limestone; it is said to be a parallel formation to beds of secondary granite. M. Brocchi, an excellent geognost, who published a memoir in the year 1811 on the valley of Fassa, has not only seen greenstone partly pyroxenic cover what he believes to be transition limestones, but which in their upper beds pass to alpine limestone with silica; he has also recognised those pyroxenic greenstones as alternating with limestone. (Melignon, Fedaja.)

I here state very extraordinary facts of position, and on which no doubt M. de Buch, who has recently visited the valley of Fassa, will throw new light. The relations of position of those countries appear very complicated. Is the rock on which greenstone and dolerites are interstratified positively zechstein, or does it belong to the transition formation? Are these greenstones and dolerites found in beds or in veins? Are the feldspathic granular rocks (called syenites and granites with three elements) oryctognostically analogous to the homonymous rocks of Christiania, or are they trachytes? Admitting that the superposition of rocks has been observed with precision, and that the various formations have been properly named we should here see repeated in secondary formations, the phenomena which MM. de Buch and Haussmann first published in the series of intermediary formations. The alternation of sedimentary, arenaceous, and crystalline rocks, would continue, by a periodical series, as far as towards the most modern formations. We already know from the excellent observations of MM. Mac Culloch and Boué, that in Scotland, and in several parts of the continent, granular, porphyritic, syenitic and pyroxenic rocks, extend from the transition formation to the coal-sandstone. Alpine limestone is immediately superposed on the formation of porphyry, and red sandstone, which is geognostically connected with that formation. From these statements it appears to me that it would not be surprising to see these very crystalline (amphibolic and feldspathic) beds, which we have already recognised in coal-sandstone, interstratified with alpine limestone. Positive geognosy ought to present a chain of facts well observed, and judiciously compared together; it does not show that the repetition of certain crystalline types stops necessarily at the coal-sandstone; the observations of Marzari would not consequently overthrow any geognostic law. If they are confirmed by future researches, they will serve to enlarge our views respecting the curious phenomenon of *alternation* in formations the most remote from each other. As veins filled with greenstone, syenites, and pyroxenic masses, traverse in several parts of the two continents primitive granite, clay-slate, transition porphyries, secondary limestones, and even formations above the chalk, several celebrated geognosts have suspected that the problematic rocks of the shores of the Avisio (Lavis) are perhaps volcanic masses or currents of lava which have proceeded from below (the interior of the earth) through crevices. This suspicion seems strengthened by the analogy of the crystalline rocks, which, it is asserted, are superposed on formations of very different ages (alpine limestone, limestone of Jura, and chalk); but the great masses of quartz which enter into the composition of the rocks, called by MM. de Marzari and Breislak *secondary granites,* seem to remove these problematic rocks from the modern productions of volcanoes. Let us hope that observations, often repeated on the spot, will dissipate every doubt. Contemptuous incredulity is as hurtful to science as a too great readiness in admitting facts incompletely studied. We must chiefly distinguish between the masses (trachytic ?) which are spread over secondary formations, and which are only superposed on them, and (amphibolic, pyroxenic, syenitic) masses with which they may be interposed. This difference of position alone may be the object of a direct observation; the problem of the origin of crystalline superposed or interposed beds belongs to geogony. Many very ancient rocks are perhaps but strata of melted matter, and the geogonic questions to which the rocks of Fassa give rise may in part be applied to porphyries and pyroxenic greenstone interposed with coal-sandstone. We must describe in every formation what it contains, and by what it is characterised. Positive geognosy stops at the knowledge of positions.

III. ARENACEOUS AND CALCAREOUS DEPOSITES (MARLY AND OOLITIC) PLACED BETWEEN THE ZECHSTEIN AND THE CHALK, AND CONNECTED WITH THOSE TWO FORMATIONS.

In ascending from the transition formation by secondary rocks to the tertiary formation, the phenomenon of *alternation* between limestone and arenaceous beds becomes more and more striking. We first see the intermediary limestone white and crystalline (Tarantaise), or compact and carburetted, alternating with grauwackes; and then succeeded by the red sandstone, alpine limestone or zechstein, variegated sandstone (red marl), muschelkalk (limestone of Gottinguen), quadersandstein (sandstone of Königstein), limestone of Jura (oolitic formation), green sandstone or secondary sandstone with lignite (green sand), chalk, tertiary sandstone with lignite (plastic clay), Parisian limestone, &c. &c. I must here call to mind six *alternations* of twelve intermediary, secondary, and tertiary formations (arenaceous and calcareous), according to their relative antiquity, as if in one spot of the earth these rocks were all simultaneously developed. By the frequent suppression of some of them, particularly of the variegated sandstone, muschelkalk, and quadersandstein, the Jura limestone (oolitic) sometimes reposes immediately on the alpine limestone (Andes of Mexico and Peru, Pyrenees, Apennines).

The deposites which we unite in this third grand

division (§§ 29—33.) form nearly the whole *of the middle sedimentary formation* of M. Brongniart. I hesitate to use denominations that relate to limits so differently traced by modern geognosts. M. Conybeare, in the excellent work he has lately published in conjunction with Mr. Phillips, on the Geology of England, distinguishes the formations into *supermedial, medial,* and *submedial.* So many systematic divisions perhaps adds to the difficulty which the synonymy of rocks already presents.

CLAY AND VARIEGATED SANDSTONE (OOLITE-SANDSTONE, SANDSTONE OF NEBRA; NEW RED SANDSTONE, AND RED MARL) WITH GYPSUM AND ROCK-SALT.

§ 29. *The sandstone of Nebra, or variegated sandstone* (Thuringia), and the red marl of England (from the banks of the Tees in Durham, as far as the southern coast of Devonshire) are not only parallel formations, that is, of the same age and occupying the same place in the series of rocks; they are identical formations. The first, poor in petrifactions (*Strombites speciosus, Pectinites fragilis, Mytulites recens, Gryphites spiratus,* Schl.), is a formation composed of three series of alternating beds; 1°, clays; 2°, micaceous and slaty sandstone, with masses of clay in flattened and lenticular forms (thongallen); 3°, oolite, generally reddish-brown. We find in the variegated sandstone of the continent, in subordinate beds, gypsum (thongyps), sometimes lamellar, most frequently fibrous, and without fetid limestone. We have seen above, that in Germany and France a great number of salt-springs flow over beds of clay and gypsum, and that at Thiede, between Wolfenbüttel and Brunswick, as well as at Sulz near Heilbronn, small masses of rock-salt are disseminated in that formation, which, at Sulz, was reached by soundings after the muschelkalk and before the zechstein. The red marl (red ground, red rock, red ford) so well examined by MM. Winch and Greenough, destitute of petrifactions and oolitic beds, and divided by fissures into rhomboidal masses, is in England the real position of rock-salt; it is composed, in its upper beds, of marly clay, gypsum (alabaster), and salt (Witton near Norwich; Droitwich); and in its inferior beds, either of conglomerates with pebbles of primitive and transition rocks, or of fine-grained sandstone (between Exeter and Axminster). The rock-salt of England, Lorraine, and Wurtemberg, connects the formation of sandstone and variegated clay, towards the lower part, with the zechstein or alpine limestone; towards the upper, in the north of Germany, this formation passes to muschelkalk, of which the most ancient beds are a little arenaceous. It may be also said, that the oolites of the variegated sandstone (Eisleben, Endeborn, Bründel) and its marls *precede* the Jura formation; but these reddish-brown oolites are insensibly lost in an arenaceous rock; they differ essentially from the white and yellowish-white oolites of the Jura limestone. On the continent the variegated sandstone is very distinct from the zechstein, notwithstanding the traces of salt by which it is allied to the latter formation ; in England, the red marl, magnesian limestone, and the conglomerates of Exeter and Teignmouth (Devonshire), which, under the name of *new red conglomerate*, represent the coal-sandstone of Mansfeld, are as closely connected together as the deposites of coal with the transition rocks (mountain limestone and old red sandstone).

In describing above the red sandstone of New Grenada, I considered the gradations of composition and structure which distinguish that coal formation from the variegated sandstone (bunte sandstein), with respect to the interposed beds of sand, slaty-clay, and coarse conglomerates. These conglomerates, which characterise the lower beds of the red marl, reappear in the chain of the Vosges. The upper beds of the variegated sandstone are green, and are supposed to be coloured by nickel and chrôme ; they are sometimes mixed with small laminæ of sulphate of barytes (Mariaspring near Gottinguen).

Subordinate beds : 1°, clay-gypsum, a little chloritic, with aragonite (Bastène near Dax), rock crystals, colourless (Langensalze, Wimmelburg), or red (Dax), and with disseminated sulphur (between Gnölbzig and Naundorf). This gypsum was formerly regarded as a particular formation placed between the variegated sandstone and muschelkalk

(Cresfeld and Helbra in Saxony, Dölau in Franconia, Neuland near Löwenberg in Silesia, Amajaque in Mexico); 2°, limestone in thin beds, sometimes marly, sometimes containing magnesia; 3°, clay impregnated with mineral pitch (Kleinscheppenstedt near Brunswick); 4°, sand (triebsand), with great chamites and petrified wood (Burgörner); 5°, very quartzose sandstone, almost without any visible cement, extremely characteristic not only of the variegated sandstone, but of the plastic clay that surrounds the strata of basalt; 6°, brown iron often in géodes; 7°, traces of coal, perhaps even of lignite, which must not be confounded with the analogous deposites of the quadersandstein, and secondary and tertiary sandstone with lignite (below and above the chalk). It is asserted, that branches of carbonised trees have been found in the clays with gypsum of Oberwiederstedt in Thuringia; and in fact, the argentiferous schists of Frankenberg (Hesse), which are nothing more than carbonised phytolites, penetrated by metals, appear to many geognosts to belong to the variegated sandstone. M. Boué, whose obliging communications have often enriched my labours, observes that the variegated sandstone exists as outliers in the south-west of France, and is there represented by marl and fibrous or compact gypsum (Cognac, S. Froult near Rochefort), and is sometimes immediately covered by Jura limestone and coarse clay. The variegated sandstone has assumed a considerable developement at the foot of the Pyrenees, between S. Giron and Rimont. As in that part of the Andes which I passed over, the secondary formations, that is, those which are above the alpine limestone, are scarcely at all developed, I think I clearly distinguished the variegated sandstone only in the following places.

At Mexico, in descending mountains composed of intermediary porphyry, and eminently metalliferous, (Real del Monte and Moran) towards the hot baths of Totonilco el Grande, a considerable formation of bluish-grey limestone occurs, almost free from shells, generally compact, but containing very white granular beds which are coarse-grained. This limestone, celebrated for its caverns (Dantö or the pierced Mountain), and traversed by veins of sulphuret of lead, appears to me to be a transition formation. It is covered by another formation that is whitish-grey and entirely compact, resembling zechstein. On the latter argillaceous sandstone reposes (bunte sandstein), the upper layers of which are (near Amajaque) clays with lamellar gypsum. I am of opinion that sandstone containing flattened masses of clay (thongallen), near Veracruz, and (Acazonica) also fine foliated gypsum, belongs, like the gypsum of Amajaque, to the variegated sandstone. Perhaps this formation of Veracruz turns round the eastern coast, and is connected with the limestone deposites of Nuevo-Leon, which are rich in argentiferous galena.

In the Llanos, or steppes of Venezuela, the clay-gypsum (Cachipo, Ortiz) is certainly posterior to the coal-sandstone; but if the limestone which separates it (between Tisnao and Calabozo), far from being zechstein, is, as its smooth fracture and its resemblance to lithographic limestone seem to indicate, of the Jura formation, these gypsums of the Llanos must be still more modern than those of the variegated sandstone. At Guire (eastern coast of Cumana) a white and granular gypsum (jurassic ?) contains large masses of sulphur. Saliferous clay mixed with gypsum and petroleum of the peninsula of Araya, opposite to the Isle de la Marguerite, is placed between zechstein and a tertiary formation. Gypsum being contained in the latter formation (hill of the chateau St. Antoine, at Cumana; plains between Turbaco and Carthagena), it might be thought that the saliferous clays of Araya are much more recent than red marl or variegated sandstone. But I dare not pronounce with confidence on the age of these formations, in the absence of so many rocks that we find elsewhere placed between the zechstein and the tertiary formation. The gypsum which I examined in the interior of New Grenada (table-land of Bogota; Chaparal at the west of Contreras (appeared to me to be of the alpine limestone formation.

When we examine the formation § 29. in countries remote from each other, we find the denomination of *variegated sandstone* as improper as the denomination of red sandstone; we may substitute for the latter that of coal-sandstone, in recollecting one of the most general and positive results of modern geognosy. It were to be wished that a geognost of great authority would substitute a geographical name for that of variegated sandstone or sandstone with brown oolite. Till then I shall continue to use the denomination of *sandstone of Nebra.*

MUSCHELKALK (SHELLY LIMESTONE; LIMESTONE OF GOTTINGUEN).

§ 30. A formation very little variable, and which the too vague denomination of *shelly limestone* has caused to be confounded, out of Germany, with the inferior or superior beds of Jura limestone (with lias, forest marble, and Portland stone). It is well characterised by its more simple structure, by the immense quantity of shells, partly broken, which it contains, and by its position above the sandstone of Nebra (bunte sandstein) and below quadersandstein, by which it is separated from the Jura limestone. It covers a vast part of the north of Germany (Hanover, Heinberg, near Gottinguen; Eichsfeld, Cobourg, Westphalia, Pyrmont, and Bielfeld), where it is more abundant than zechstein or alpine limestone. It extends, in the south of Germany, over the whole table-land between Hanau and Stutgard. In France, where notwithstanding the great and useful labours of M. Omalius d'Halloy, the secondary formations below the chalk have been so long neglected, MM. de Beaumont and Boué recognised it around the whole chain of the Vosges. Muschelkalk has generally pale tints, whitish, greyish, or yellowish, of a dull and compact fracture; but the mixture of small laminæ of calcareous spar, produced perhaps by the remains of petrifactions, renders it sometimes a little granular and brilliant; several beds are marly, arenaceous, or passing to the oolitic structure (Seeberg near Gotha; Weper near Gottinguen, Preussisch-Minden, Hildesheim). Hornstone, passing to gun-flint and jasper (Dransfeld, Kandern, Saarbrück), is either disseminated by nodules in muschelkalk, or there form small beds little continued. The lower beds of this formation alternate with the variegated sandstone (between Bennstedt and Kelme), or are connected insensibly with sandstone, being charged with sand, clay, and even (at the east of Cobourg) with magnesia (magnesiferous beds of muschelkalk).

Subordinate beds. Marl and clay, so frequent in the Jura limestone, variegated sandstone, and zechstein, are somewhat rare in the muschelkalk. This rock contains, in Germany, hydrate of iron, a little fibrous gypsum (Sulzbourg, near Naumbourg), and coal (lettenkohle of Voigt; at Mattstedt and Eckardsberg near Weimar) mixed with aluminous slate and carbonated fruits (coniferes?). The more the coal advances towards the tertiary formations, the nearer it approaches, at least in some of its beds, to the state of lignite and aluminous earth.

Petrifactions. According to the researches of M. de Schlottheim, and rejecting the beds that do not belong to the muschelkalk: *Chamites striatus, Belemnites paxillosus, Ammonites amalteus, A. nodosus, A. angulatus, A. papyraceus, Nautilites binodatus, Buccinites gregarius, Trochilites lævis, Turbinites cerithius, Myacites ventricosus, Pectinites reticulatus, Ostracites spondyloides, Terebratulites fragilis, T. vulgaris, Gryphites cymbium, G. suillus, Mytulites socialis, Pentacrinites vulgaris, Encrinites liliiformis,* &c. Some insulated beds of Jura limestone contain perhaps still more petrifactions than the muschelkalk; but the remains of organised bodies abound in no secondary formation so uniformly as in that which we have just described. An immense quantity of shells partly broken and partly in good preservation, but adhering strongly to the stony matter (entrochites, turbinites, strombites, mytulites), are accumulated in several strata from 20 to 25 millimeters thick, which traverse muschelkalk. Many species are found united by families (belemnites, terebratulites, chamites). Between these eminently shelly strata are distributed ammonites, turbinites, some terebratulites, of which the shells have still their nacre, the *Gryphæa cymbium,* and very fine pentacrinites. Corals, echinites, and pectinites are rare. The abundance of entrochi in muschelkalk has occasioned them to give to that formation, in some parts of Germany, the name of *limestone with entrochi* (trochitenkalk). As a bed of entrochi often characterises zechstein also, and separates it from the coal-sandstone, this denomination may occasion two very distinct formations to be confounded together. The denomination of *limestone with gryphites* (graphytenkalk of zechstein and of the Jura limestone), and all those which bear allusion to fossil bodies, without indicating the species, expose us to the same danger. It is asserted that muschelkalk contains the bones of large animals (oviparous quadrupeds? Freiesleben, t. i. p. 74., t. iv. p. 24. 305.) and birds (ornitholites of Heimberg; Blumenbach, *Naturgesch*, t. iii. *Aufl.* p. 663.); but these bones may probably belong, like the teeth of fish, to the breccias or marls superposed on muschelkalk?

Some celebrated English geognosts, MM. Buckland and Conybeare, thought they recognised, in their travels through Germany, the muschelkalk of Werner as identical with the lias, which is the lower bed of the Jura limestone. I am inclined to think that, notwithstanding the bluish-grey oolites observed in the muschelkalk on the banks of the Weser, there is rather a parallelism than an identity of formation. The muschelkalk occupies the same place as the lias; it abounds equally in ammonites, terebratulites, and encrinites; but the species of fossils differ, and its structure is much more simple and uniform. The strata of muschelkalk are not separated by the blue clays that abound in the upper and lower layers of the lias formation. The middle beds of that formation have a smooth dull fracture, and resemble much more the lithographic varieties of Jura limestone than the muschelkalk of Gottinguen, Jena, and Eichsfeld. M. d'Aubuisson thinks that this latter formation is represented in England by the Portland stone, cornbrash, and forest marble; but whatever analogy these beds of marly limestone with shells partly broken (forest marble) may present, we must recollect that they alternate with formations altogether oolitic, and are separated from the red marl by the lias, exactly as oolitic Jura limestone is separated by muschelkalk from the variegated sandstone. M. Boué recognised muschelkalk in France, in the table-land of Burgundy, near Viteaux and Coussy-les-Forges, and near Dax in the commune of S. Pan de Lon, &c. I did not observe it in the equinoxial part of America. The very arenaceous beds, filled with madrepores and bivalve shells, of the coast of Cumana and Carthagena, which I formerly thought were connected with them, are probably tertiary formations.

QUADERSANDSTEIN (SANDSTONE OF KÖNIGSTEIN).

§ 31. A very distinct formation (banks of the Elbe, above Dresden, between Pirna, Schandau and Königstein; between Nuremberg and Weissenberg; Staffelstein in Franconia; Heuscheune, Adersbach; Teufels-mauer at the foot of Hartz; valley of the Moselle, and near Luxembourg; Vic in Lorraine; Nalzen in the province of Foy, and Navarreins, at the foot of the Pyrenees), characterised by M. Haussmann, and long confounded either with the quartzose varieties of the variegated sandstone, and the sandstone of the plastic clay (trappsandstein), or with the sandstone of Fontainebleau, above the calcaire grossier of Paris; it is the white sandstone of M. de Bonnard, the sandstone of third formation of M. d'Aubuisson. Preferring geographical denominations, I often name this formation, *sandstone of Königstein;* the variegated sandstone, *sandstone of Nebra;* and the muschelkalk, the *limestone of Gottinguen.*

The quadersandstein is of a whitish, yellowish, or greyish colour, very fine-grained, with an argillaceous or quartzose cement, almost invisible. Mica is there little abundant, always silvery, and disseminated in insulated spangles. It is destitute both of those interposed beds of oolite, and of those flattened or lenticular masses of clay (thongallen) which characterise the variegated sandstone. It is never slaty, but divided into beds little inclined, very thick, cut at right angles by fissures, and some of which easily decompose into very fine sand. It contains hydrate of iron (Metz) disposed in nodules. The organic remains disseminated in this formation afford, according to MM. de Schlottheim, Haussmann, and Raumer, an extraordinary mixture of pelagic shells, very analogous to those of the muschelkalk and dicotyledon phytolites. There have been found in it mytulites, tellinites, pectinites, turritellæ, oysters (no ammonites, but cerithia; Habelschwerd, Alt-Lomnitz in Silesia), and at the same time wood of the palm-tree, and impressions of leaves belonging to the class of dicotyledons, and small deposites of coal (Deister, Wefersleben near Quedlinbourg), well described by MM. Rettberg and Schulze, passing to lignite. These vestiges of wood, of a bituminous quality, may no doubt surprise us in a formation so far removed from the great formation of lignite which is placed between the chalk and coarse Parisian limestone; but recent observations show us traces of true lignite as far as in the limestone with gryphæa arcuata below the lias (le Vay, coast of Caen), and as far as the variegated sandstone. The coal of inferior quality in the muschelkalk, and consequently of a more ancient formation than the quadersandstein, passes also to lignite.

M. de Raumer had also observed that the quadersandstein is separated from the variegated sandstone by muschelkalk (limestone of Gottinguen); it is placed between this limestone and that of Jura, and consequently below the great oolitic formations of England and the continent. We cannot consider it in this position, with M. Keferstein (see his interesting Essay on the Mineralogical Geography of Germany, t. i. p. 12. & 48.), as parallel to the molasse of Argovia (mergelsandstein), which represents the plastic clay (tertiary sandstone with lignite) above the chalk. The nature of the vegetable remains contained in the quadersandstein, and its connection with the planerkalk, which belongs to the chloritic and arenaceous beds of the chalk, has led several celebrated geognosts to regard it as a formation posterior to Jura limestone; thus, it is placed by MM. Buckland, Conybeare, and Philipps, between the chalk and the last oolitic beds. But, according to the observations of M. Boué and several other celebrated geognosts of Germany, the quadersandstein (sandstone of Königstein), alternating sometimes with marly beds and conglomerates, reposes immediately on gneiss near Freiberg; on coal-sandstone in Silesia and Bohemia; on variegated sandstone (sandstone of Nebra) near Nuremberg in Franconia; on muschelkalk (limestone of Gottinguen), Hildesheim and Dickholzen near Helmstadt, and near Schweinfurt on the Mein. It is covered by Jura limestone, and alternates with the marly beds of that limestone, in Westphalia, between Osnabrück, Bielfeld, and Bückebourg.

JURA LIMESTONE (LIAS, MARLS, AND THE GREAT OOLITIC DEPOSITES).

§ 32. A very complex formation, composed of alternating beds of marly and oolitic limestone, containing gypsum and a little sandstone. The mode of partial alternations, though very constant in every locality, varies in countries of considerable extent; we recognise, however, on the most distant points of Europe, a striking analogy between the great divisions or principal beds. In the series of formations, the nearest of secondary rocks, the limestone of Jura (*Jurassus*) is placed between the quadersandstein and the chalk. The latter even passes into it insensibly, and may often be considered, by the analogy of its fossils, as a continuation of the Jura limestone. The superposition of this limestone on the quadersandstein, so long contested, is seen in Germany, according to M. Schmitz, near Wilsbourg; according to M. Boué, near Blumenroth Staffelstein, and between Osnabruck and Bückebourg. When the three formations of quadersandstein, muschelkalk, and variegated sandstone, are not simultaneously developed, the Jura limestone, by the suppression of the intermediary members of the geognostic series, immediately covers the zechstein, or alpine limestone. In that case (northern declivity of the Pyrenees; Apennines, between Fossombrono, Furli, and Nocera; Cordilleras of Mexico, between Zumpango and Tepecuacuilco), we see the latter pass insensibly to a whitish limestone, with an even dull fracture (or conchoidal with very flattened cavities), which cannot be distinguished from the compact beds of the Jura limestone destitute of oolites. This passage, with which M. Charpentier was also struck in the south of France, merits a very attentive examination. Notwithstanding the great difference which exists between the fossil remains of the muschelkalk and Jura limestone the last secondary formations are closely connected together; and we must not be surprised that in a series α, β, γ, δ, ϵ, the formation α (zechstein) makes a passage to ϵ (Jura limestone), on account of the frequent suppression of the terms β, γ, δ (that is, of the variegated sandstone, muschelkalk, and quadersandstein). The arenaceous formations, β and δ, alternate with clay and marl more or less abundant, so that by a great developement of their disaggregated beds these reduce the stony strata to the state of simple interposed beds, and end, as it happens in the west of France, by filling up the whole interval between α and ε.

Jura limestone covers without interruption a great extent of country, from the chain of the Alps as far as the centre of Germany, from Geneva as far as Streitberg and Muggendorf in Franconia. This formation containing caverns towards the north with fossil bones, has singularly arrested the attention of the German geognosts. M. Werner considered it as identical with the muschelkalk. I observed since the year 1795, that it differs from it essentially, and I proposed to designate it by the name of limestone of Jura, on account of the perfect analogy that exists between the western mountains of Switzerland and those of Franconia. This denomination is now generally received; but it has been proved that the Jura limestone, instead of being placed below the variegated sandstone (as I had erroneously believed with the greatest number of geognosts, confounding this sandstone with the molasse of Argovia and the sandstone of Dondorf and Misselgau near Bareuth), is more recent than the variegated sandstone, muschelkalk (Bindloch), and quadersandstein (Schwandorf, Phantaisie (?), and Nuremberg). This interposition between the quadersandstein and the chalk, which is founded on direct observations, explains very well the gradual passage (mountain of S. Pierre near Maestricht), of the *craie tuffeau* to the Jura formation. The name often given to the latter of cavernous limestone (höhlenkalk) may lead to erroneous comparisons. We must distinguish between formations of which the whole mass is cellular and full of cavities, and rocks with caverns. Several of these rocks containing vast caverns are neither porous nor cellular. The transition limestone (mountain limestone of Derbyshire) would deserve in England and Hartz, almost as much as that of Jura, the appellation of *cavern limestone*. Rauchkalk and rauchwacke, on the contrary, which form the middle layers of the zechstein in Thuringia, and which were erroneously thought to be parallel to the Jura limestone, are, like the latter, full of small cavities from 2 to 10 lines of diameter in a very considerable extent of beds, without having on that account any real caverns. The phenomenon of caves and that of the *porosity* of the mass, are not necessarily united; they are modifications which, far from characterising any particular formation, are found in formations that are very different from each other.

Although the partial beds that compose the Jura limestone on the continent are very unequally developed, and that the order of their succession often varies, we constantly remark a certain number of distinct beds spread over a very considerable extent of country. We shall name them here, beginning by the most ancient; marly limestone (and very hard calcareous marls), greyish-blue, analogous to the lias of England (according to MM. Boué and Buckland, *Essai géogn. sur l'Ecosse*, p. 201., and *Struct. of the Alps*, p. 17.), sometimes traversed by veins of calc-spar filled with the gryphæa arcuata, yellowish-grey oolites alternating with marl partly bituminous, and with gypsum; compact limestone with a smooth and dull fracture, and white oolites; beds filled with madrepores analogous to the coral limestone of Normandy and the coral rag of England; slaty limestone with fish and crustacea (Pappenheim and Solenhoffen). The lower bed of this complex formation is particularly marked in France (Burgundy), and in the south of Germany (Wurtemberg), by the name of *limestone with gryphites;* but some geognosts seem inclined to separate this bed from the limestone of Jura, considering it with MM. de Buch and Brongniart, as belonging to the zechstein, or with M. Keferstein, as being parallel to the muschelkalk. Here an important question arises, that of knowing in what relation of position and composition the limestone with gryphites of Jura is found, with that which bears the same name in the north of Germany, and which M. Voigt has made known since the year 1792? A great analogy between the nearest beds of two formations sometimes immediately superposed to each other has nothing in it surprising; the same species of gryphæa may occur in very distinct formations still more remote from each other, but the geognostic connection observed between limestone with the gryphæa arcuata, alternating with marl, and the other lower beds of the Jura, make me lean to the opinion that this limestone, and that of the gryphæa with spines (gryphitenkalk of Voigt), placed beneath the variegated sandstone, are not of the same formation. M. Mérian, in his excellent Monography of the vicinity of Bale, states also this opinion, and regards, like M. Haussmann, the argillaceous sandstone of Rheinfelden, on which the Jura limestone reposes, as the variegated sandstone; while M. de Buch (Mérian, *Umgeb. von Basel*, p. 110.) takes it for the coal-sandstone, and supposes that, by the non-developement of the variegated sandstone, the oolitic and lithographic beds of the Jura repose in that locality immediately on the gryphite beds that belong to the zechstein. I have considered it as my duty in this essay to state the opinions of the most celebrated geognosts, even when they are opposite to my own.

It is indubitable, and what we think useful again to mention, that the Jura limestone which near Laufenbourg reposes on granite, at Schwarzwald on red or coal sandstone, and near Geneva on alpine limestone, is in the centre of Germany placed on the quadersandstein. The superposition of a rock on the latest formation determines its place as a term of the geognostic series. We generally see in Franconia and the Upper Palatinate only the upper beds of Jura limestone, which are the most compact. Marl and oolites are here much more rare than in the west of Switzerland, and in France (Caen, Lons-le-Saulnier). According to M. de Schmitz, we find between Eichstädt and Ratisbon, a limestone very cellular; granular beds containing druses filled with sand; compact and conchoidal limestone with nodules of flint; slaty, and fissile limestone, analogous to that of Sohlenhofen and the lithographic slabs of Heuberg near Kolbingen. These beds full of cavities (valley of Laber near Berodhausen, Pegnitz, Creussen, Tumbach), which I found also in Italy (valley of Brenta between Carpane and Primolano), at the isle of Cuba (between Potrero de Jaruco and the port of Batabano), at Mexico (table-land of Chilpansingo), give to the surface of the country, which is covered with small pointed rocks, a very peculiar aspect.

In the west of France, according to M. Boué, an uninterrupted band of Jura limestone extends from S.E. to N.W., from Narbonne and Montpellier as far as la Rochelle, separating towards the north the transition formation of La Vendée, and the primitive formation of Limousin. The marly and oolitic beds have assumed a much greater developement on the coast of Normandy than in Germany. We shall state, from the interesting researches of M. Prévost, the superposed beds between Dieppe and the Cotentin, beginning, as usual, by the most ancient beds; 1°, limestone with gryphæa arcuata and lithographic limestone

(Le Vay, Issigny), containing lignite, and superposed on the transition formation; 2°, inferior beds of clay and oolites (clay of the Vaches-noires, alternating with lias containing remains of the ichthyosaurus; the grey oolites of Dive, which are ferruginеous, mixed with clay containing lignites, and numerous petrifactions of madrepores, modiola, *Gryphæa cimbium* and ammonites; white oolites); 3°, limestone of Caen; the lower beds with nodules of silex, and few shells (ammonites, belemnites), and having some bones of the crocodile; the upper beds contain madrepores (coral-rag), trigoniæ, and cerithia, perfectly analogous to those found above the chalk; 4°, upper beds of clay at Cape la Hève, of a bluish colour, with lignite, remains of crocodiles (Honfleur) and calcareous beds, less developed than at Caen. We see, that in this part of Europe lignite extends through all the beds of Jura limestone, and that this formation, deducting the interposed clay, is composed of three great beds, viz. limestone with the gryphæa arcuata, oolite, and limestone with madrepores and trigoniæ.

In England, the formation of Jura, stretching without interruption from Yorkshire to Dorsetshire, fills the whole space between the red marl (variegated sandstone) and the chalk; for we know no formation between the Jura limestone and red marl of a composition analogous to the muschelkalk and quadersandstein; two rocks also often wanting on the continent. The English and Scotch geognosts, who have lately studied the structure of their country with indefatigable zeal, distinguish the beds of Jura limestone by denominations partly very characteristic, and several of which resemble the subdivisions acknowledged on the continent; 1°, *Lias* with a little flint, covering the saliferous red marl, analogous to the limestone with the gryphæa arcuata of the continent; the upper two thirds consist of an argillaceous bluish mass, alternating with beds of limestone; towards the lower part these beds increase in thickness, become white, and pass to a stone fit for lithographic purposes (bones of the ichthyosaurus, and near twenty species of ammonites, belemnites). 2°, *Lower system of oolites*, viz. oolite mixed with sand, fullers' earth, great bed of oolite (great oolite) with remains of shells; oolite-slate of Stonesfield, forest marble, cornbrash, and kelloway rock, shelly and arenaceous limestone. 3°, *Middle system of oolites*, viz.; Oxford clay (clunch-clay of M. Smith), sand, and calcareous conglomerates (calcareous grit), coral-rag, or limestone with madrepores and echinites. 4°, *Upper system of oolites;* clay of Kimmeridge, blue, a little bituminous, analogous to the blue clay of Cape la Hève in Normandy, which are also above the limestone with madrepores and oolites; Portland stone with ammonites; Purbeck stone, an argillaceous limestone filled with shells, alternating with marl and gypsum. I have followed the divisions of MM. Smith, Conybeare, and Philipps, which differ a little from those adopted by M. Buckland. The three systems of oolite in England are separated by formations of clay. With respect to the oolitic structure itself, we have already observed that traces of it are found in very different formations; there are some beds of oolite, according to MM. de Gruner and Escher (*Alpina*, t. iv. p. 369.), in the transition limestone of Switzerland, in the coal-sandstone (Freiesleben, *Kupfersch*, b. iv. p. 123.), in the alpine limestone or zechstein (Hartlepool in Northumberland) in the variegated sandstone (Thuringia, Vic in Lorraine), and in the muschelkalk.

Subordinate beds: hornstein (flint) in small continuous beds; magnesian limestone (Nice); fetid limestone and gypsum, with traces of rock-salt (Kandern, see Merian, *Umgeb, von Basel*, p. 36.); slaty and micaceous sandstone, sometimes siliceous, interstratified in layers with gryphites (Hemmiken, Waldburgstuhl, Lons-le-Saulnier); globuliform oxid of iron (bohnenerz), in the Jura limestone (Neufchâtel), Frickthal, Wartenberg in Swabia), and between this limestone and the molasse or tertiary sandstone with lignite (Arau, Baden); coal, with impressions of fern (?) and mixed with pyrites (Neue Welt, Bretzweil).

Petrifactions: the Jura limestone is, after the formations above the chalk, that of which the fossil remains have been best determined in England, France, and the west of Switzerland. It contains, like the still more ancient formations, quadersandstein and zechstein with copper-slate, pelagic shells, mixed with lignite and bones of fresh-water saurian animals, and even, if we are not mistaken in the zoological determination, with bones of the didelphus (Stonesfield slate). I am not certain whether the mixture of marine and fluviatile shells, so evident in the greater part of the tertiary formations, has been observed in the formations below the chalk. Where the Jura formation is almost destitute of marl and oolites (Franconia, Upper Palatinate; Carniole, between S. Sesanne and Triest), very considerable beds are entirely without petrifactions. The remains of oviparous quadrupeds, fish, and tortoises, are found in almost all the beds in the most recent (Purbeck stone) as well as in the most ancient (lias); the latter, however, contain the most, and it appears that they have only the ichthyosaurus (proteosaurus of Sir Everard Home) and the plesiosaurus, which is an analogous animal, and not the real crocodile. This difference in the distribution of reptiles has been also observed by M. Prévost on the western coast of France. The bones of the ichthyosaurus are found (principally?) in the limestone beds (lias) which belong to the beds of clay below the oolite, while crocodiles are only found above the oolite. In England, according to MM. Smith, Conybeare and Philipps, among the prodigious number of petrified shells of which the genera are not known, the following are distinguished; *Ammonites giganteus*, *A. excavatus*, *A. Duncani*, *A. Banksii*, *A. angulatus*, *A. Grenoughi*, *Nautilus striatus*, *N. truncatus*, *Trochus dimidiatus*, *T. bicarinatus*, *Trigonia costata*, *T. clavellata*, *Terebratula intermedia*, *T. spinosa*, *T. digona*,

Ostrea gregaria, O. palmata, Modiola lævis, M. depressa, M. minima, Pentacrinites caput Medusæ, P. basaltiformis, etc. Although the species of ammonites (twenty in number), belemnites, and pentacrinites, described as in lias, are not identical with those of the muschelkalk, it appears to me very remarkable that we find three families in rocks of nearly the same age, between the last strata of zechstein (alpine limestone) and the first or most ancient of the Jura limestone. MM. Prévost, Lamouroux, and Brongniart will soon enrich zoological geognosy by the profound researches they have made on the shells and zoophytes found on the coast of France, between Dieppe and the Cotentin, in Franche-Comté, and in Switzerland. We shall in the meantime content ourselves with mentioning here the fossil bodies which occur in the Jura limestone of the continent, from Geneva as far as Franconia, according to notes which I took from a catalogue of M. de Schlottheim: *Chamites jurensis, Belemnites giganteus, Ammonites planulatus, A. natrix, A. comprimatus, A. discus, A. Bucklandi, Myacites radiatus, Tellinites solenoides, Donacites hemicardius, Pectinites articulatus, P. æquivalvis, P. lens, Ostracites gryphæatus, O. cristagalli, Terebratulites lacunosus, T. radiatus, Gryphites arcuatus, Mytulites modiolatus, Echinites orificiatus, E. miliaris, Asteriacites pannulatus,* Turritellæ, Hippuritæ (*Cornucopiæ* of Cape Passaro in Sicily), *Gryphites arcuatus, etc.* It is deserving of attention that this gryphæa arcuata which M. Sowerby calls *Gryphites incurvus*, and which characterises the lower strata of the Jura formation in Switzerland and on the eastern coast of France, is also the shell, which after the *Ammonites Bucklandi* and the *Plagiostoma gigantea*, characterises most the lias in England. The beds of white and granular limestone that are found frequently in this formation (Neufchâtel, Monte Baldo) are owing to petrifactions of madrepores.

We have already seen fish in greater or less abundance, but belonging to very distinct genera, in transition clay-slate (Glaris), in the carburetted slate of the red sandstone (Goldlauter and Allthal, near Kleinschmalkalden), in the alpine limestone and its copper marl, and even in the muschelkalk, (very rarely, Esperstedt, Obhaussen): these ichthyolites are more frequent in the Jura limestone, particularly in its upper beds. From thence they ascend above the chalk into the tertiary sandstone with lignite (plastic clay), into the calcaire grossier (Monte Bolca), gypsum with bones (Montmartre), and fresh-water limestone (Œningen). I mention the formations that present analogous phenomena, according to the order of their relative age, to prevent the errors that arise from ignorance of those analogies.

A justly esteemed geognost, M. Buckland, is inclined to regard the fissile limestones of Pappenheim and Sohlenhofen, celebrated for their impressions of fish and crustacea, as superposed on the Jura limestone, and belonging to the calcaire grossier of tertiary formation; these slaty limetones appear to me, on the contrary, altogether analogous to the Purbeck stone of England, which abounds also in petrifications of fish, and forms, like the limestone of Pappenheim, the latest bed of the Jura formation. I had occasion, in 1795, to examine the fine quarries of Sohlenhofen, conjointly with M. Schöpf, and in going from Muggendorf by Ansbach to Pappenheim, we observed a close connection between the various beds of the same formation. MM. de Buch, Boué, and Beudant are of this opinion respecting the ichthyolites of Franconia.

Both Jura limestone and calcaire grossier exist in the Vicentine, and both contain madrepores. In my first travels, however, in Italy (1795), I thought that the long bands of ramose corals which traverse, in forming veins (between l'hôtellerie du Monte di Diavolo and lake Fimon, at the west of Lungara), the summit of Monte di Pietra nera, belong rather to the Jura limestone, perhaps to the strata called in England coral-rag. These bands of madrepores, which have remained in their natural place, are two feet broad; they present a very singular aspect, and pass through limestone masses almost free from petrifactions in a regular direction N. 80° E., rising like a wall above the surface of the soil. M. Boué has also observed an analogous phenomenon in the Jura limestone (coral-rag), which surrounds the basin of Vienna, and the lower strata of which contains nagelfluhe, analogous to the *calcareous grit* of the oolitic formation of England (Filey in Yorkshire).

I thought I recognised in the equinoxial zone of America the Jura formation in several whitish limestones, partly lithographic, with a fracture smooth and dull, or very flat conchoidal. These are the limestones of the cavern of Caripe (south-east of Cumana), the shore of Nueva Barcelona (Venezuela), the isle of Cuba (between the Havannah and Batabano; between Trinidad and the boca del Rio Guaurabo), and the central mountains of Mexico (plains of Salamanca and defile of Batas). The white limestone of Caripe, which resembles perfectly that of the caverns of Gailenreuth in Franconia, is superposed in the bluish-grey alpine limestone of Cumanacoa, The Jura formation of the shore of Nueva Barcelona contains small beds of hornstone, passing to a black siliceous schist (a phenomenon which occurs also near Zacatecas at Mexico); it is covered (Aguas calientes del Bergantin) like the alpine limestone at the summit of the Impossible, with a very quartzose sandstone. It might be imagined that this sandstone of Bergantin belongs to the quartzose beds of green sandstone, or secondary sandstone with lignite; but, as it also forms beds in the alpine limestone (Tumiriquiri), it remains very doubtful if the sandstone of Bergantin and of Tumiriquiri are different formations, or if beds quite similar extend from the alpine limestone into the Jura formation. This formation abounds less than any other secondary formation in arenaceous rocks. We have however mentioned above, beds of sandstone in the western mountains of Switzerland, at Waldburgstuhl, Eptigen, and Hemmiken near Bale. In the vast steppes of Venezuela, near Tisnao, it appeared to me that the red sandstone immediately supports (as at Schwarzwald in Swabia) a lithographic limestone very analogous to the Jura limestone. This position is also to be found at Mexico, in the plains of Temascatio, south-west of Guanaxuato. At the northern extremity of the valley of Mexico (between Hacienda del Salto, Batas, and Puerto de Reyes), a greyish-blue limestone formation with a smooth fracture, containing gypsum, and supporting a limestone-breccia, appeared to me to belong to the Jura formation, notwithstanding the proximity of tertiary marls (Desague of Huehuetoque), in which fossil bones of elephants are buried. I might also mention the passage which is observed from the alpine limestone to a limestone entirely similar to that of Arau and Pappenheim, at the western declivity of the Cordilleras of Mexico, between Sopilote, Mescala, and the rich mines of Tehuilotepec; but in that region the Jura formation is less marked than at the isle of Cuba, the islets of Cayman, and in the mountains of Caripe near to Cumana. In the part of the new world through which I passed, I no where saw the variegated sandstone, muschelkalk, or quadersandstein, separate alpine limestone from the formations which I have just described. Destitute of oolite, they also abound little in petrifactions of shells, and beds of marl. Their dull and smooth fracture gives them altogether the aspect of the Jura limestone of Germany and Switzerland. Are these limestone formations of America, the Pyrenees, and the Apennines, which appear so closely connected with the alpine limestone (zeschtein), only the newest beds of the latter, and ought they to be separated from the real Jura limestone, rich in shells, oolite, and marl? This important question can be resolved only by multiplying the observations of position, which are much more decisive than those of composition and exterior aspect.

FERRUGINOUS SAND AND SANDSTONE, GREEN SAND AND SANDSTONE, SECONDARY SANDSTONE WITH LIGNITE (IRON SAND AND GREEN SAND).

§ 33. This division contains sandstone and sand with lignite, placed *below the chalk*; they are two arenaceous formations, coloured by iron, separated by a bed of clay, (weald clay), and superposed on Jura limestone, (oolite formation). In England they attain the thickness of a thousand feet, and are found in the west of France, where MM. Prevost and Boué have made them the object of a profound study.

The yellowish-brown *ferruginous sands* alternate with siliceous sandstone and small masses of iron ore, often worked with advantage; it contains fossil wood and lignite (Bedfordshire, Dorsetshire).

The *green sands*, coloured by a protoxide of iron, alternate with calcareous and siliceous sandstone, with conglomerates, yellowish marl with crystals of gypsum, and even with small beds of compact limestone, which have been sometimes confounded with the Portland stone. We find there nodules of hornstone and calcedony (Sarlat in Périgord), small deposites of hydrate of iron, a resin that passes to amber (isle d'Aix near Rochelle; Obora and Alstadt in Moravia), and a great number of fossil remains, several of which (*cidaris*, *spatanges*) resemble those of the chalk. The siliceous sandstone of this formation contains impressions of dicotyledon leaves. The green sand towards the upper part passes to a chalky marl (chalk-marl of Surrey). The green or chlorite earth, which characterises the bed of sand nearest the chalk, is found in formations of very different age, in the coal-sandstone of Hungary (on the frontiers of Galicia), in the variegated sandstone and gypsum which belongs to it, and in the quadersandstein and lower beds of the calcaire grossier of Paris. According to the excellent researches of M. Berthier on the green grains of the chalk and calcaire grossier, these grains are a silicate of iron; but it is probable that the quantities of magnesia and potash vary in different formations, as they vary, according to the analysis of Klaproth and Vauquelin, in the green earth of Verona (talc-chlorite zoographique of Haüy) and in earthy chlorite. The analogy that sometimes occurs between the quadersandstein of Germany and the siliceous beds of the green sandstone (iron sand) either in their solid state, or in a state of disaggregation, has led several geognosts to confound those two formations. M. Boué, who has so well explored the positions of the rocks in Scotland, England, and Germany, recognised green sandstone (exactly similar to that of the vicinity of Oxford) in France, along the Mayenne and the Loir, from la Ferté-Bernard beyond la Flèche, in the department of Charente, in Mans, Saintonge, and Périgord.

To this same formation, § 33., the lignites also of the isle of Aix belong, on which M. Fleuriau de Bellevue has made such interesting researches. According to that learned geologist, the sub-marine forest on the coast of Rochelle consists of flattened dicotyledon wood, partly petrified, partly bituminous or fragile, and sometimes passing to a state of jet. These woods are penetrated by pyrites, and pierced by the teredo and marine worms. The holes resulting from this perforation are filled with quartz agathe, and sulphuret of iron. The trees are found either in horizontal beds, sometimes in a parallel direction, sometimes collected in disorder. The wood, when altogether or partly petrified, reposes on a green sand; those which are in a fibrous or bituminous slate repose on beds of plastic clay of a deep blue. They are surrounded by marine algæ and small branches of lignite. Among the masses of algæ is found a resin that passes to amber; it is friable and of various colours. Trunks of trees heaped together form a band of a league and a half in breadth, from the extremity of the north-west of the isle of Oleron as far as fourteen leagues in the interior of the continent, on the right bank of the Charente. This band is more than seven feet thick; it runs from W.N.W. to E.S.E., and is three feet above the level of the sea at low tide. Where the lignites are covered by the ocean they are incorporated (like the masses of succin-asphalte and the great bones of marine animals) with a coarse sandstone which reposes on plastic clay. The position of these deposites is from below upwards (according to an unpublished memoir of M. Fleurian de Bellevue): 1°, compact limestone (lithographic), with a smooth fracture (La Rochelle, S. Jean d'Angely); 2°, oolite beds (point of Chatelaillon and Matha); 3°, lumachelle and beds of madrepores with impressions of *Gryphæa angustata*; (these three beds constitute the Jura formation, of which the bed of madrepores represents the coral-rag); 4°, a great bed of lignite with marine turf, succin-asphalte, and plastic clay; 5°, ferrugineous and chlorite sand, slaty clay; arenaceous and calcareous beds with trigoniæ and cerithia; fragments of lignites. At the south-west of the Charente, where the beds Nos. 4. and 5. are wanting, horizontal beds of very white limestone repose immediately on the oolites of the Jura formation, and represent the lower strata of the chalk. M. Boué has seen traces of these lignites stretching from Rochefort by Perigueux as far as Sarlat.

These sands and clays with the lignites of the

green sandstone are connected towards the lower part with the blue clay with lignites of Cape la Hève (near Havre); above, they precede in a manner the great deposite of lignite of tertiary formation, that is, the lignite of the plastic clay and molasse which occurs above the chalk. The chalk itself contains lignite in its lower strata (chlorite-chalk, between Fécamp and Dives), and may in some respects be regarded as a continuation of the Jura formation; the phenomena which we have just mentioned are well worthy of the attention of geognosts. The *plänerkalk* of Germany, often mixed with mica and grains of quartz, forms one of the upper layers of the green sandstone, representing at the same time chlorite-chalk and a part of the coarse and tuffaceous chalk.

IV. CHALK.

§ 34. We have seen that in proportion as we are removed from the alpine limestone the formations become more complex. The muschelkalk and quadersandstein have indeed a simple structure, but the Jura limestone and green sandstone, where they are well developed, present a great number of beds and frequent alternations. This tendency to a varied composition, to a grouping of heterogeneous masses (a tendency which attains its maximum in the tertiary formations), diminishes in some degree at the chalk formation. Placed between the green sandstone and the plastic clay, or the sandstone with tertiary lignite, the chalk, by a great simplicity of structure, is contrasted with the complex formations we have just named. The argillaceous beds (*dief*), and the calcareous and arenaceous beds (*tourtia*) which separate the Jura formation (oolite) from that of the chalk, ought not to be confounded with the latter formation, although it is often not easy to fix the limits between the marl with oolitic beds of the Jura formation, the strata of green sandstone, and those argillaceous marls or yellowish and almost compact limestones which seem to belong to the lower beds of chalk.

The latter formation is composed, according to the researches of MM. Omalius and Brongniart, of three beds sufficiently distinct. The lower is *chlorited chalk* (*glauconie crayeuse*), friable, and with green disseminated grains; the middle is the *craie tufau*, or *coarse chalk*, greyish, sandy, containing marl, and, instead of gun-flint (silex pyromaque), hornstone of a light colour. The upper strata is *white chalk*. Sometimes the most ancient beds assume a darkish-grey colour, and become very compact (vicinity of Rochefort), or are granular and friable (mountain of St. Pierre near Maestricht). The chlorited chalk often passes insensibly into green sand. The white chalk is the purest of the limestone beds of different ages; it contains only one or two per cent. of magnesia, but it is mixed with a greater or less quantity of sand. The connection between the chalk formation of Paris and the other secondary formations (between Gueret and Hirson) has been shown in a section by M. Omalius (Bull.

phil., 1814). M. Gay-Lussac and myself, in a barometric levelling made in 1805, from Paris to Naples, saw appearing successively below the chalk Jura limestone, alpine limestone, red sandstone, gneiss, and granite (between Lucy-le-Bois, Avallon, Autun, and mountain of Aussy). The chalk formation which was too long neglected, extends much farther than is generally believed. It has been observed in several parts of Germany, for instance, in Holstein, Westphalia (from Unna to Paderborn), in Hanover, at the foot of the Hartz near Goslar, in Brandenburg near Prentzlow, and at the isle of Rugen. It can often be only recognised by the fossil bodies which exhibit fragments of marly and arenaceous formations. It contains few heterogeneous beds, for instance, beds of clay (Isle of Wight; Anzin); flint, either in plates or nodules, in straight lines or in small veins (Isle of Thanet; Brighton), and characterising the upper strata of chalk. Here also we find globular pyrites and sulphate of strontian (Meudon).

Petrifactions. In the basin of the Seine, according to the observations of MM. Defrance and Brongniart, there are found in the upper beds of the chalk many belemnites (*Belemnites mucronatus*), and echinites (*Ananchites ovata*, *A. pustulosa*, *Galerites vulgaris*, *Spatangus coranguinun*, *S. bufo*), oysters (*Ostrea vesicularis*, *O. serrata*), terebratulites (*Terebratula Defrancii*, *T. plicatilis*, *T. alata*), pectens (*Pecten cretosus*, *P. quinque-costatus*), *Catillus Cuvieri*, *Alcyonium*, astéries, millepores, &c. The chalk, distinguished by M. Brongniart as craie tufau and glauconeuse, contains (vicinity of Havre, Rouen, and Honfleur, Perte du Rhône near Bellegarde;) *Gryphea columba, G. auricularis, G. aquila, Podopsis truncata, P. striata, Terebratula semiglobosa, T. gallina, Pecten intextus, P. asper, Ostrea carinata, O. pectinata, Cerithium excavatum,* trigonies, crassatelles, encrinites and pentacrinites (England), and which is very remarkable, nautilites and several ammonites (*Nautilus simplex, Ammonites varians, A. Beudanti, A. Coupei, A. inflatus, A. Gentoni, A. rhotomagensis*), while the upper beds of the chalk, near Paris, do not contain (with the exception of *Trochus Basteroti*) a single univalve shell with a simple and regular spire. According to the researches of MM. Buckland, Webster, Greenough, Philipps, and Mantell, compared with those of M. Brongniart, the greatest analogy exists between the organic remains found in France and England in the strata of the chalk of the same age. Every where, the most ancient beds contain bones of great saurians (monitor) and of sea-turtles, teeth and the vertebræ of fish (squales). Notwithstanding the analogies which the sandstone with lignite exhibits (green sand and plastic clay), below and above the chalk, this formation belongs rather to the secondary than to the tertiary formation, to which several celebrated geognosts refer it. According to M. Brongniart, the shells of the argillaceous formation approach much nearer those of the Jura formation than the shells of the calcaire grossier, from which the chalk is geognostically separated in the most distinct manner.

TERTIARY FORMATIONS.

The considerations which I have stated above, on the intimate connection between the last beds of the transition formations, and the first of the secondary formations, may in great part be applied to the connection observed between the secondary and the tertiary formations. The transition rocks are however, more closely connected with the coal-sandstone, than the chalk is to the formations by which it is succeeded. What is most important in geognosy, is to distinguish well the partial formations, and not to confound what nature has clearly separated, and to assign to each term of the geognostic series its true relative position. With respect to the attempts that have been recently made to unite several of these formations by groups and sections, they have had the fate of all *generalisations* differently graduated. The opinions of geognosts have remained more divided with respect to the great than to the small divisions. The same *formations* have almost every where been admitted, but the nomenclature of the groups which should unite them has varied. Thus, botanists are more agreed on the determination of the genera than on the subdivision of the same genera, among neighbouring families. I preferred preserving in the tabular arrangement of formations the ancient classifications which have been generally received. In this long series of rocks, this assemblage of monuments of different epochas, we distinguish chiefly three very striking phenomena; the first dawn of organic life on the globe, the appearance of fragmentary rocks, and the catastrophe which has buried the ancient monocotyledon vegetation. These phenomena mark the epocha of intermediary rocks, and that of the coal-sandstone, first member of the secondary rocks. Notwithstanding the importance of the phenomena which we have just remarked, the rocks of one epocha have always some prototype in the rocks of a preceding epocha, and every thing denotes the effect of a continued developement.

As the names, *formations of middle sediment*, *new alpine limestone*, &c., are employed in many modern geognostic works, without always specifying the rocks that are contained in those groups, it will here be useful to describe the synonymy of this nomenclature of positions. M. Brongniart, distinguishing between *primitive* and *primordial*, comprehends with M. Omalius d'Halloy, under the denomination of *primordial formations*, all the *primitive* and *intermediary* crystalline rocks of the school of Freiberg; he divides his secondary formations (Flötzgebirge) into three classes. In the first, that of the *lower sediment* (*Descr. géol. des environs de Paris*, p. 8.; *Sur le gisement des ophiolithes*, p. 36.), are comprehended mountain or transition limestone, the red sandstone or coal-sandstone, the alpine limestone or zechstein, and the lias; in the second, that of the *middle sediment*, Jura limestone and the chalk; in the third, that of *upper sediment*, all the beds that are newer than the chalk. The formation of *upper sediment* replaces consequently the *tertiary formation*, a denomination quite as improper for designating a *fourth* formation succeeding to the primitive, intermediary, and secondary formations, as were the ancient names of *terrains à couches* (secondary rocks) and *terrains à filons* (primitive and transition rocks). M. de Bonnard, in his interesting *Aperçu géognostique des formations*, excludes from *primordial formations*, porphyries, transition syenites, and all crystalline rocks posterior to those which contain remains of organised bodies; he regards (and we prefer his method) the word *primordial* as synonymous with *primitive*. The *upper secondary formations* of M. Bonnard differ much from the *formation of upper sediment* of M. Brongniart; they are rather those which that geognost calls *formation of middle sediment*. All the formations from the chalk to the red sandstone, with the exception of the coal, is comprised in the supermedial order of M. Conybeare, while the close connection observed in England between the deposites of coal and the rocks by which they are supported, have induced M. Buckland (*Structure of the Alps*, 1821, p. 8. 17.) to extend the secondary formations from the chalk as far as the mountain limestone and grauwacke (old red sandstone). He considers our zechstein with saliferous deposites as the *elder alpine limestone;* and the lias, oolites, green sand and chalk, as the *newer alpine limestone.* These indications will, I believe, suffice for the understanding of the synonymy of the great geognostic divisions.

The frequent mixture of stony beds, loose materials or disintegrated masses, had long occasioned the tertiary formations, that is, those which are posterior to chalk, to be confounded with the *alluvial deposites* which Guettard had called (in 1746) the *zone of sand.* The tertiary formations have been erroneously considered as of small importance, irregular in their stratification, and confined to small districts. The school of Freiberg at first (1805) placed only few formations above the muschelkalk and the chalk, viz. the sands and clays with lignite, observed by Hollmann in 1760 (*Phil. Trans.* vol. li. p. 505.); calcareous nagelfluhe, travertino, and fresh-water tufa. (Reuss, *Geogn.*, t. ii. p. 473. 630. 644.) Bruguiére had already observed that the meulière of Montmorency contained only fresh-water shells. The gypsum with bones of Montmartre, which Karsten considered as analogous to the saliferous gypsum of zechstein, had been considered by Lamanon and M. Voigt (1790) as a deposite from fresh water. Werner regarded it as altogether different from the gypsum formations of Germany, and as being of a much more recent epocha. (Freiesleben, *Kupfersch.*, t. i. p. 174.) The observations collected by the *Geological Society* of London, and the *Wernerian Society* of Edinburgh, the useful travels of M. Omalius d'Halloy (1808), and of some Italian geognosts, had furnished a considerable mass of materials for the study of the tertiary formations; but we can only date a more precise knowledge of the different beds that constitute this formation and present the same characters in the most remote countries, from the time when the *Geological Description of the Environs of Paris, by MM. Brongniart and Cuvier*, appeared (1st edition, 1810; 2d edition, 1822). All the tertiary formations (with the exception perhaps of the sandstone with lignite which belongs to the plastic clay) are found best developed in the basin that surrounds that capital. All those that are wanting in other parts of Europe, or found only as outliers, are united on the banks of the Seine.

In characterising succinctly the terms of the *tertiary series*, I shall avail myself both of the great work of M. Brongniart, of that which has just appeared by MM. Conybeare and Phillips on the geology of England, the geological travels of M. Beudant in Hungary, and the recent observations of MM. Boué and Prevost, who, in filling up the void between the tertiary and oolitic formations, have performed a great service to positive geognosy. It is by the comparison of formations very distant from each other, that we can avoid to a certain degree confounding the geognosy of positions with the geographical description of an insulated basin. It is remarkable enough that the last strata of the great geognostic edifice, that in which the epocha of formations draws nearest to our own times, has been examined so lately. As the disaggregated masses of the tertiary formation contain fossil shells in a high degree of preservation, this formation has also given rise to the improvement of subterranean conchyliology. The predilection for this science, which prevails in several countries, will become equally useful to the study of secondary and intermediary formations, if the combination of the zoological characters with those of the position and relative age of rocks be not neglected.

I have already stated the motives which induced me to avoid the denominations of *first*, *second*, and *third marine formation*, or *fresh-water formation*. I have most frequently substituted geographical names to those numerical denominations, which are very apt to give rise to erroneous ideas. The most recent formations are those of which the positions appear to have been the most modified by local circumstances. A periodical alternation of calcareous and siliceous substances (clay itself contains near 70 per cent. of silica) is manifest even in the strata which belong to the same formation. In some countries the heterogeneous beds, and subdivisions of calcareous or gypseous formations assume such a thickness, that they appear like particular or independent formations. It thence results, that the *succession* and *parallelism* of the tertiary rocks, which are so recent and of so complex a structure, may sometimes differ from the type we assign to them in our tabular arrangement of formations.

TERTIARY CLAY AND SANDSTONE WITH LIGNITES (PLASTIC CLAY, MOLASSE, AND NAGELFLUHE, OF ARGOVIA).

§ 35. At the beginning of the tertiary formation, as well as below chalk, or between that rock and the Jura limestone, we find deposites of lignite; thus, we have seen the great deposite of *coal* (coal measures), placed on the limit of the intermediary and secondary formations. The two secondary and tertiary formations commence by masses of buried plants. As we advance from the coal-sandstone towards the more recent formations, the monocotyledon plants are by degrees replaced by dicotyledon plants; the former are even found above the chalk, as far as in the gypsum with bones (endogenites of M. Adolphe Brongniart, but not ferus); in general, however, dicotyledons (exogenites) predominate in the deposites of lignite. I am less surprised at this mixture than at the uniformity of the monocotyledon vegetation of the ancient world, of which we see the remains in the intermediary formations, and in the coal-sandstone. Amidst the forests of Oroonoko, which are extremely rich in monocotyledons, their proportion to that of dicotyledons is, with respect to the mass, that is, to the number of plants, as one to forty. The proportion therefore which the coal formations present is not *tropical;* has it been modified by the unequal resistance which the monocotyledons and dicotyledons oppose to destruction?

We shall unite in this formation *of sandstone with lignite above the chalk*, the parallel formations of plastic clay, marl, and sand with lignites, molasse, and nagelfluhe.

There is only a fragment of this formation in the vicinities of London and Paris; it is found much more developed in the south of France, Switzerland, and Hungary. The chalk is covered in France and England by a bed of plastic clay, without shells and organic remains, entirely destitute of lime, and containing some silex and selenite. A bed of sand separates the plastic clay from beds of clay (*fausses glaises*), which are more siliceous and darker. The latter contain lignite or fossil bituminous wood produced by monocotyledon and dicotyledon plants, real amber (according to the discovery of M. Bequerel), bitumen, and (Soissonnois, Montrouge, Bagneux) a mixture of pelagic and fluviatile shells (cyrenæ, fresh-water cerithia or potamides, melaniæ, lymneæ, and paludinæ). This mixture is observed usually only at the upper limit of the plastic clay and lignite. According to M. Prevost, the sea-shells resemble those of the calcaire grossier. Interposed beds, sand and sandstone with shells, masses of concretioned limestone, with crystals of sulphate of strontian.

The fossils, according to MM. de Ferussac and Brongniart, are: *Planorbis rotundatus, Paludina virgula, P. unicolor, Melanopsis buccinoidea, Nerita globulosa, Melania triticea,* — *Ceritium funatum, Ampullaria depressa, Ostrea bellovaca, &c.* In England, the plastic clay, which must not be confounded with the *London clay* (representing the calcaire grossier of Paris), nor with the *Oxford,* or *Clunch clay* (of the Jura formation), abounds more in sand than clay; it contains lignite (Isle of Wight, Newhaven), and, what is remarkable on account of the analogy of this formation with the molasse of Argovia and Hungary, a friable sandstone (Stutland in Dorsetshire). In this formation, according to MM. Webster and Buckland, impressions of leaves, fruits of the palm-tree, cyclades (*Cyclas cuneiformis, C. deperdita*), turritellæ, cerithia (*Ceritium melanoides, C. intermedium*), and oysters have been found (*Ostrea pulchra, O. tenuis*).

The *formation containing amber,* of Pomerania, and Prussia, probably placed on the chalk, is composed of clay, lignite, and nodules of amber. The organised bodies it contains have been recently examined by M. Schweigger. From its position, as M. Brongniart has judiciously observed, it belongs to the formation § 35.

Sandstone with lignite (molasse and macigno) are spread over the plains of Hungary, as well as in the great basin of Switzerland, between the Alps and the Jura, or rather between the lake of Annecy, and that of Constance. The formation of Hungary, which M. Beudant has made known, is geognostically the most important, because it is superposed on Jura limestone (Sari Sap, in the vicinity of Gran, and banks of the lake Balaton). It is immediately covered (near Buda) with shelly limestone analogous to the limestone of Paris. It is composed of pudding-stones (nagelfluhe) and calcareous breccias alternating with a micaceous sandstone that is friable, slaty, with small angular grains of quartz, sand, and beds of clay. It contains considerable deposites of lignite (Csolnok, south of Gran, Wandorf near Œdenbourg), springs of bitumen, granular hydrate of iron, fresh-water shells, and, at the contact with the superposed calcaire grossier, salt-water shells. The *arenaceous formation* of Switzerland, which comprehends the molasse and nagelfluhe, is composed, according to the late researches of MM. de Charpentier and Lardy (beginning by the lower beds), 1°, of sandy *limestone*, a little ferruginous, passing often to a real sandstone with a calcareous cement; 2°, pudding-stone (*nagelfluhe*) containing calcareous and siliceous fragments, constantly round and agglutinated by a calcareous cement; 3°, molasse or sandstone with small grains of quartz, and with an argillaceous and marly cement. Veins of calcareous spar often traverse the nagelfluhe and the molasse (fine and friable sandstone), and alternate with beds of marl. The nagelfluhe, which contains both pebbles of porphyry and of compact limestone (Rigi, Fribourg, Entlibuch), is not always covered by molasse; and M. de Buch has long since remarked, that between Habkern and little Emmethal, molasse alternates several times with nagelfluhe. The whole of this formation, the surface of which is generally uncovered, is placed immediately towards the north (Arau, Porentruy, Boudry) on Jura limestone, and towards the south on alpine limestone (vicinity of Geneva and Teufenbachtobel south-east of Rigi). Some celebrated geognosts have long regarded nagelfluhe, from the inclination of the beds, as anterior to the alpine limestone. M. Keferstein also considers the molasse (mergelsandstein) to be below the chalk, and even Jura limestone. A fetid and bituminous limestone, a fibrous and clay gypsum, alternating with marl which contains *ammonites*, a compact yellowish-brown limestone, and lignite, form the subordinate beds of the molasse in Switzerland. The deposite of lignite which is worked near S. Saphorin, between Vevay and Lausanne, is covered by nagelfluhe; that of Paudex is interposed in the molasse. The whole of this formation contains in Switzerland marine shells (ammonites, cythereæ, donax) and also fresh-water shells (lymneæ, planorbes), palmacites with flabelliform leaves (Montrepos), and bones of quadrupeds (Aarberg, Estavayer, Kæpfnach on the banks of the lake of Zuric), which, according to the researches of M. Meisner, belong to the *Anaplotherium*, the *Mastodon angustidens*, and the *Beaver*. In the molasse of Cremin and Combremont, a shelly marine breccia reposes on a brown limestone filled with planorbes. M. Brongniart, from the year 1817, insisted on the analogy of the plastic clay of Paris, with a part of the formation of nagelfluhe, and the molasse of Switzerland so long confounded with the variegated sandstone of Germany. He also thinks that the molasse which contains bones of mastodontes and of *anthracoterium* (Cadibona near Savone), is still more recent than the plastic clay; that it is either connected with the calcaire grossier, which is often arenaceous, or is the parallel of the gypsum of Montmartre. The bones of vertebrated animals, rarely found in the plastic clay of Paris and London (near Auteuil and Margate), have not yet been zoologically determined, and M. Cuvier, in pursuing his important researches on the position of fossils, has not hitherto recognised any remains of *land-mammiferæ*, except in formations posterior to the calcaire grossier. From these considerations, the molasse or sandstone with lignite of Hungary may have been anterior to those of Switzerland; but, as the formations of calcaire grossier (Parisian limestone) and gypsum with bones are scarcely developed in the latter country, and that in general the frequent alternation of the tertiary rocks renders their parallelism a little uncertain, the long period of the formation of molasse and nagelfluhe, in Switzerland (that of lower, upper, arenaceous, marly, calcareous, and gypseous beds), may have been contemporary with the three formations of the plastic clay, calcaire grossier, and gypsum of the vicinity of Paris.

The formation with which we are now occupied is, according to the recent observations of M. Boué, extremely developed in the south-west of France, from Libourne to Agen, above all, at the north of the Dordogne and the Gironde, where it reposes on chalk. It is there composed (beginning by the upper beds) of calcareous sandstone, filled with the remains of shells, and the bones of vertebrated animals, small beds of globular iron, grey and greenish marl, and yellowish limestones with cerithia. Deposites of lignite have there been recognised by M. Brongniart (*Descr. géol.*, art. ii. § 1.), but they are not numerous; and the position of this arenaceous formation between the chalk and the calcaire grossier of Bourdeaux characterises it sufficiently as the molasse. The sandstone with lignite may locally be destitute of lignite, as the red or coal-sandstone is often destitute of coal. As almost all the secondary formations have *their sandstone* and *their conglomerates,* we must not consider all the nagelfluhes of Europe (polygenic puddingstones of the classification of M. Brongniart) as belonging to the same formation, § 35. There are some which appear to be only local formations, and of small extent; others (Salzbourg and S.Gall?), according to the judicious observation of M. Boué, are perhaps more ancient than the chalk and Jura limestone. The analogy also of some beds placed between the quadersandstein and the chalk with those placed between the chalk and gypsum with bones, is a phenomenon well worthy of the attention of geognosts.

Immense deposites of sand, clay, and lignite with mellite (Artern), and with amber (bernstein of Muskau, and bernerde of Zittau), cover a part of Germany. Beds of very quartzose sandstone are there found (Carlsbad, Habichtswald, Meissner, Wilhelmshöhe near Cassel, Wolfseck), chiefly where beds of basalt are superposed on clay with lignite. On account of this proximity, the improper denomination given anciently to this sandstone (which may be mineralogically confounded with the quartzose part of the variegated sandstone, and with that of Fontainebleau), was that of trap-sandstone (*trapp-sandstein*). Do the sands with garnets (granatensand), that is, the clays and marl of Meronitz and Podsedlitz in Bohemia, which contain disseminated pyropes, belong to the same formation, § 35., or, as several phenomena observed in the Cordillera of Mexico, and the Isle of Graciosa (Archipelago of the Canaries), would lead me to suppose, to basaltic clay of igneous formation?

LIMESTONE OF PARIS (CALCAIRE GROSSIER, OR CALCAIRE A CERITES), THE PARALLEL FORMATION TO THE LONDON CLAY, AND THE ARENACEOUS LIMESTONE OF BOGNOR.

§ 36. This very complicated formation, found in Hungary, in Italy, and in the new continent, was entirely misunderstood before the publication of the *Mineralogical Geography of the Environs of Paris.* The calcaire grossier, separated by a bed of plastic clay, consists, in the basin of the Seine, according to M. Brongniart, of thin beds alternating very regularly, of limestone more or less hard, and argillaceous or calcareous marl. The fossil shells are generally the same in corresponding beds through a very considerable extent of country, and exhibit a remarkable difference in the species in different systems of beds. This phenomenon of uniformity in the distribution of animals chiefly characterises the tertiary formation; we begin to observe it in the different beds that compose the Jura formation in Switzerland and England. The lower beds of the calcaire grossier of Paris are chloritous (glauconeuses), arenaceous, and contain madrepores and nummulites. In the middle beds many impressions are found of leaves and stalks of plants (*Endogenites echinatus, Flabellites parisiensis, Pinus Defrancii,* according to the work of M. Adolphe Brongniart on fossil vegetables), milliolites, ovulites, cythereæ, but scarcely any cerithia. The upper beds contain lucinæ, ampullaria, striated corbulæ, and a great variety (nearly sixty species) of cerithia; but, in general, the latter strata abounds less in fossil bodies than the middle and lower strata, in which MM. Defrance and Brongniart have collected near 600 species of shells. The famous shelly bed of Grignon and the fossils of the *Falun de Touraine* belong principally to the middle strata. In these, and in the system of the upper beds, the calcareous layers are sometimes entirely replaced by sandstone or masses of hornstone (hornstein). MM. Gillet de Laumont and Beudant observed (between Pierrelaie and Franconville near Beauchamp) that these sandstones exhibited a mixture of marine and fresh-water shells (limneæ and paludinæ). The fossils of the Parisian limestone, among which belemnites, orthoceratites, baculites, or ammonites are never found, differ entirely from those of the chalk.

Some of the shelly deposites which represent, in different parts of Europe, the formation we describe, are identical in composition and aspect (plains of Vienna, described by M. Prevost; hills of Pest and Teteny in Hungary, described by M. Beudant), but they are sometimes analogous only by their geognostic position, and by the fossil remains they contain (England). The calcaire grossier of Hungary containing cerithia, turritellæ, ampullaria, venuses, and crassatellæ, scarcely recognisable, because only the mould remains, present the most minute empyrical characters by which the Parisian limestone is recognised. They are connected with shelly sands (Czerhat, Raab) which are partly mixed with green grains, and which have a great analogy with the shelly deposites of the plains of Lombardy.

The calcaire grossier of the Dordogne and the Gironde, which are geographically nearer to the basin of the Seine, does not always display that resemblance of composition which we have just remarked in those of Hungary. It is composed, according to the recent observations of M. Boué, of two very distinct strata. The lower is a little shelly, with broken fossil bodies, and contains compact yellowish-white limestone sometimes marking like chalk, marl, and beds of quartzose pebbles. The upper strata is a sandy limestone, extremely shelly, and sometimes nearly resembling brownish molasse.

In England, according to the researches of MM. Buckland, Webster, and Sowerby, the London clay is not only by its superposition to plastic clay a *parallel formation* to the limestone of Paris, but also contains almost all the species of shells which seem to belong more particularly to the lower beds of that limestone. The formation in the basin of the Thames, which the English geognosts designate commonly by the name of *London clay*, is only a deposite of clay and brownish marl, containing sulphuret of iron and some crystals of selenite; but this bed, in other parts of England, approaches much nearer to the calcaire grossier, in its mineralogical composition. It presents, according to MM. Conybeare and Phillips, on the coast of Sussex, at Bognor, and near Harwich (Essex), beds of compact and sandy limestone, in which are found, besides the fossil bodies peculiar to the analogous formation in the basin of Paris, impressions of fish, bones of turtles and crocodiles (Islington), a species of ammonites (*ammonites acutus*, at Minstercliff) and lignite. The *Ceritheum giganteum*, common in the clay of London, belongs in

France only to the lower bed of calcaire grossier, which is also destitute of every other species of cerithia. The *London clay*, in which amber has been found (Holderness in Yorkshire), appears to have a nearer relation with the plastic clay (tertiary sandstone with lignite) than the calcaire grossier of Paris.

M. Brongniart connects with this formation the greater part of the calcareo-trap formations of the Vicentine (Val Ronca, Montecchio maggiore, Monte Bolca), the hill of the Supergue of Turin, the cape S. Hospice near Nice, the Great Land of Guadeloupe, &c. The celebrated impressions of fish of Monte Bolca, on which M. Blainville has undertaken an interesting work, are not properly found in the calcaire grossier, according to the researches of M. Maraschini, but (which is seen particularly at Novale and Lugo near Salceo) in a fetid and slaty limestone, separated from coarse sandstone by a bed of clay with lignite. This position appears to me to connect the bituminous marl (of Monte Bolca) with the impressions of fish and leaves, to the marl of the gypsum with bones of Montmartre.

In equinoxial America, where I did not observe any formations of chalk and sandstone with lignite, the hills that border on some parts of the Cordillera of Venezuela, on the sea-side (Castillo of San Antonio of Cumana, Cerro del Barigon in the peninsula of Araya, Vigia de la Popa near the port of Carthagena), appear to me to belong to the calcaire grossier. Those hills are composed, 1°, of a *compact and arenaceous limestone*, whitish-grey, of which the beds, sometimes horizontal, sometimes irregularly inclined, are five or six inches thick (some beds are nearly destitute of petrifactions, others are filled with madrepores, cardites ostracites, and turbinites, mixed with large grains of quartz); 2°, a *calcareous sandstone*, in which grains of sand are more frequent than shells (several layers of this sandstone, containing, not spangles of mica, but nodules of brown iron, and being so siliceous that they scarcely effervesce with acids, fossil bodies being wanting in them altogether); 3°, *beds of hardened clay* with selenite. The calcareous stratum, of which I deposited considerable specimens in the cabinet of natural history at Madrid, exhibit (between Punta Gorda and the ruins of the castle of Santiago d'Araya) an innumerable quantity of solens, ampullaria, oysters, and corals, partly disposed by families. This tertiary formation, composed of shelly limestone, with grains of quartz, clay, marl, and calcareous sandstone, is found geographically connected with the tertiary formations of the islands opposite to the coast of Cumana, for instance, Guadelope and Martinico. It sometimes reposes immediately on the alpine limestone (Punta Delgada), sometimes on the saliferous clay of Araya, of which I have spoken above (§ 28. p. 321).

SILICEOUS LIMESTONE, AND GYPSUM WITH BONES, ALTERNATING WITH MARL (GYPSUM OF MONTMARTRE).

§ 37. According to the principles of classification which I have followed in this work, I might have separated the siliceous limestone (calcaire siliceux, Champigny) from the gypsum alternating with marls both marine and of fresh-water; but, not having been able, in the course of my travels, to make the formation above the chalk a particular object of my studies, I do not wish to change any thing in the general section shown in the work of MM. Brongniart and Cuvier.

The *siliceous limestone* of the basin of Paris, which is sometimes tender and white, sometimes greyish, very fine-grained and cavernous, is, as it were, penetrated through all its mass by silex, or quartzose matter. It is closely connected, towards the upper part, with gypsum, by the argillaceous and gypseous marl which alternates both with siliceous limestone and gypsum with bones (ridge of the Briffe of S. Denys, Crecy, Coulommiers); and, towards the lower part, with the calcaire grossier, the last beds of which exhibit sometimes, also, siliceous infiltrations; but the hornstone of the calcaire grossier contains marine shells, while the siliceous limestone of the gypsum formation, used for millstones, has fluviatile shells in its upper beds. I have already observed above (§ 28. p. 331.) that on the back of the Cordilleras, at the height of 1800 toises, a very ancient calcareous formation (alpine limestone) affords the same curious phenomenon of a siliceous infiltration. Analogous modifications in the composition of rocks, and the chemical mixture of substances, have taken place at very different epochas. The calcareous marl which alternates with the siliceous limestone of Paris, contains *magnesite*, first made known by MM. Brongniart and Berthier, and which is a silicate of hydrate of magnesia almost pure. The siliceous infiltrations of this formation pass sometimes to a calcedony divided into laminæ and to hornstone coloured red, violet, and brown.

In the basin of Paris the gypsum formation is composed of alternating beds of slaty marl and saccharoïd gypsum compact and foliated. It contains in its middle part, and where its mass is greatest, land productions; but towards its upper and lower limits, both in the gypsum and marl, it contains marine productions. The lowest strata of the gypsum formation is characterised by the silex menilite and large crystals of lenticular and yellowish selenite. The beds of marl become fewer towards the middle, where sulphate of strontian and skeletons of fish are more particularly found. The upper division is characterised by a multitude of bones of land mammiferæ, now unknown on the globe (*Palæotherium crassum, P. medium, P. magnum, P. latum, P. curtum, Anaplotherium commune, A. secundarium, A. marinum,* the *Chæropotame* and the *Adapis* of M. Cuvier); by bones of birds, crocodiles, tryonix, and fresh-water fish: it is covered by beds of calcareous and argillaceous marl, some containing wood of the palm-tree, planorbes, limneæ and cythereæ (*Cytherea elegans*); others cerithia (*Cerithium plicatum*, *C. cinctum*), venuses, and large thick oysters (*Ostrea hippopus*, *O. pseudochama*, *O. longirostris*, *O. cyatula*). Towards the upper limit of the gypsum formation, a bed of green marl separates the fresh-water from the pelagic shells. Towards the lower part, the gypsum itself furnishes marine fossils (No. 26. of the third mass of Montmartre). Sometimes this formation is not entirely developed; gypsum is wanting, and its place is only recognised by green marl accompanied by strontian. The gypsum with bones having as yet been studied only in very few places (basin of Paris, Puy-en-Vélay, Aix en Provence), the characters that we attribute to this formation so important to geogony, or the history of the ancient revolutions of our planet, are probably not sufficiently general.

SANDSTONE AND SAND ABOVE THE GYPSUM WITH BONES (SANDSTONE OF FONTAINEBLEAU).

§ 38. This formation consists of two divisions; the lower is without shells; the upper contains marine shells. Siliceous sands and sandstone form very thick and extended beds, the surfaces of which are not parallel. In the bed without shells *in situ* (those of Villers-Cotterets and Thury appear to M. Brongniart to be worn as if they had been rolled), a quantity of scales of mica occur in some places, also nodules of brown iron disposed in beds, a little gypsum, and argillaceous marl and infiltrations of carbonate of lime (Forest of Fontainebleau). The upper division, which contains marine shells (*Oliva mitreola, Cerithium cristatum, C. lamellosum, Corbula rugosa, Ostrea flabellula*), sometimes passes to an arenaceous limestone (Romainville, Montmartre). The immense tertiary formation of Italy, that of the *subapennine hills*, with bones of cetaceæ and *Ostrea hippopus*, which extends from Asti in Piedmont as far as Monteleone in Calabria, and which M. Brocchi has described so well, belong, for the most part, according to the discussions of MM. Prevost and Brongniart, to the sandstone and sand which repose on the gypsum of Montmartre.

FRESH-WATER FORMATION, WITH CAVERNOUS MILLSTONE (MEULIERE), ABOVE THE SANDSTONE OF FONTAINEBLEAU (CALCAIRE À LYMNÉES).

§. 39. This is the great upper fresh-water formation, composed in some parts of argillo-ferruginous sand, marl, and siliceous meuliere filled with cavities (with shells, plateau of Montmorency; without shells, La Ferté-sous-Jouarre); in other places, of silex, marl, and compact limestone (Chateau-Landon). These limestones contain potamides, lymneæ, planorbes, bulimi, helices, and many impressions of plants (*Culmites anomalus, Lycopodites squam-*

matus, Chara medicaginula, Nymphæa Arethusæ of M. A. Brongniart). We refer our readers, for the history of this great fresh-water formation, which has already been found in almost every part of Europe, to the 2d edition of the *Description Géologique des Environs de Paris* (art. 8.).

One part of the globe, where the greater part of the tertiary formations have acquired a considerable developement, and where, for this reason, these formations have remained sufficiently distinct, has served as a type in the tabular arrangement of the tertiary formations; but we must not forget that in other countries this developement stops at the plastic clay, or at the calcaire grossier: the gypsum of Montmartre and the sandstone of Fontainebleau, therefore, appear to be indicated only by the parts occupied by marl and sand. The tertiary formation includes beds that are confounded together where they have not attained an equal thickness, and where the frequent alternation of marl tends to mark the limits of different strata. It might remain for me to speak of *alluvial deposites*, which present important problems on the origin of sand in desarts and steppes (produced by the red sandstone, variegated sandstone, quadersandstein, and tertiary formation?); but these deposites, so varied in their alternation, cannot be the object of a work on the superposition of rocks.

VOLCANIC FORMATIONS.

FOR the reasons which I have mentioned above, I have made secondary and volcanic formations succeed, as by a kind of bisection, to the intermediary formation (Uebergangsgebirge). This arrangement has the advantage of bringing the transition porphyries and syenites, with their porous, pyroxenic, and interposed beds (§§ 23. and 24., Holmstrand in Norway, Andes of Popayan, Cordilleras of Mexico), nearer on one hand to the porphyries, amygdaloids, and dolerites of the red sandstone (§ 26., Noyant and Figeac in France, Scotland), and on the other, to the trachytes, phonolites, and basalts of the formation exclusively called pyrogenic. In a table of position, much is already gained, when we do not separate what is connected in nature by real geognostic affinities.

We may consider the group of rocks that are generally united in a volcanic formation in a double point of view, according to a certain conformity observed in their position and superposition, or according to the relation of their composition and common origin. In the first case, without opposing the mode of formation of trachytes and basalts to that of primitive and intermediary formations, we observe the place, which, as terms of the geognostic series, ought to be occupied by the great systems of rocks composed of feldspar, pyroxene, hornblende, olivine and titaneous iron, found at the north and south of the equator uncovered, in circumstances altogether analogous, and as if superadded to more ancient formations. This manner of investigating and classing volcanic rocks is the most conformable to the progress of positive geognosy. Trachytic and basaltic rocks are united, not according to their mineralogical composition, and the apparent conformity of their origin, but according to their grouping and position; they are distributed among other rocks from their relative age, as has been done in the primitive and intermediary formations of granular limestone (§§ 10. and 20.), euphotides (§§ 19. and 26.), and porphyries (§§ 18. 22, 23. and 26.). In the second case, by the denomination of volcanic formation, every thing is insulated which is considered as incontestably of igneous origin; the terms of the pyrogenic series are opposed to other series of rocks which are said to be of an *aqueous origin.* Thus, what nature presents by gradual passages is separated in an absolute manner; and, instead of exploring the position, or placing rocks in the order of their succession, a preference is given to historical questions on the mode of their formation.

I confess (and we cannot be too explicit in the first foundations of a science) that these classifications, according to the various hypotheses which we form on the *origin of things*, appear to me not only vague and arbitrary, but also very injurious to the progress of *geognosy of position;* they prejudge in an arbitrary, and too absolute a manner, what is at least extremely doubtful. In dividing the formations, according to a superannuated custom, into *primitive*, *intermediary*, *secondary*, *tertiary*, and *volcanic*, they admit, in some degree, of a double principle of division; that of the relative age or succession of formations, and that of their origin. If we distinguish between *masses of lava* and *rocks*, or between *volcanic rocks*, rocks of *neptunian formation*, and substances formed by a pretended *aquoso-igneous liquefaction*, we tacitly attribute to granites, porphyries, and intermediary syenites, to the dolerites, and amygdaloids of the red sandstone, a mode of formation diametrically opposite to that of igneous fusion. According to this manner of proceeding, which rather belongs to *geogony* than to *positive geognosy*, all that is not comprehended in the *volcanic rocks*, in the rocks of trachyte and basalt which cover the other formations, is considered as formed in the *humid way*, or as precipitated from an *aqueous solution*. It is almost useless, in the actual state of physical science, to remind the reader how little the hypothesis of an aqueous solution is applicable to granite and gneiss, to porphyry and syenite, to euphotides and jasper. I will not here venture to decide on the circumstances which may have accompanied the first formation of the oxidated crust of our planet; but I do not hesitate to range myself on the side of those geognosts who rather conceive the formation of crystalline siliceous rocks by fire than by an aqueous solution in the manner of travertino and other fresh-water formations. Besides, the words *lava* and *volcanic rocks* are as vague as the word volcano, which sometimes designates a mountain terminated by an *ignivome* mouth, sometimes the subterraneous cause of every *volcanic phenomenon.* The trachytes which rise on the back of the Cordilleras belong indubitably to pyrogenic rocks, and yet the mode of their formation is not that of currents of lava posterior to the hollowing of the valleys. The action of volcanic fire by an insulated cone, by the crater of a modern volcano, differs necessarily from the action of that fire across the ancient fissured crust of our planet.

In considering volcanic phenomena in the most general manner, and in collecting what has been observed in different parts of the globe, we see these phenomena differing from each other, even in our days, in the most striking manner. The volcanoes of the Mediterranean, which alone have been carefully studied, cannot serve as a type to the geognost, and afford him the solution of the great geogonic problems. The absolute height of ignivomous mouths, varying from 100 to 2950 toises (Stromboli and Cotopaxi), has not only an influence on the frequency of eruptions, but also modifies the nature of the ejected masses. Some volcanoes act no longer but by their flanks, although they have still a crater at their summit (Peak of Teneriffe); some have lateral eruptions (which I found at Antisana in the Andes of Quito, 2140 toises high), without their summit having ever been pierced; others, hollow within, as several phenomena indicate (trachytic dome of Chimborazo, 3350 toises), have no permanent opening at the summit or on their flank (Yana-Urcu, a small cone of eruption situated in the table-land of Calpi), and it may be said, act only *dynamically* in disturbing the surrounding earth, fracturing the beds, and changing the surface. Rucu-Pichincha (2490 toises), which was the particular object of my researches, has never ejected a current of lava since the hollowing of the actual valleys, nor has Capac-Urcu (near Riobamba nuevo), which before the overthrow of its summit was loftier than Chimborazo. The great Mexican volcano of Popocatepetl (2771 toises), has, on the contrary, had overflowings of lava in the form of narrow bands, like the small volcanoes of Auvergne and the south of Italy. The islands that issue (almost periodically in some latitudes) from the bottom of the sea, are not, as it has often been erroneously said, heaps of scoria similar to Monte novo de Pouzzole, but heaved up rocky masses, and in which the crater opens only after their elevation. (*Relat. histor. de mon Voyage aux régions équin.*, t. i. p. 171.; and *Essai politique*, t. i. p. 254.) In the interior of Mexico, on a trachytic table-land more than thirty-six leagues distant from the sea, and far from every burning volcano, mountains 1600 feet high issued from a fissure (29th September, 1759), and threw out lavas which contain granitic fragments. All around, a surface of four miles square was heaved up in the form of a bladder, and thousands of small cones (hornitos de Jorullo), composed of clay and balls of basalt in concentric layers, are scattered on this rounded surface. All the burning volcanoes and peaks of New Spain, that rise above the limit of perpetual snows, are found on a narrow zone (*parallel to the great heights*, between 18° 59′ and 19° 12′ of latitude), which is perpendicular o the great chain of mountains. It is like a fissure of 137 leagues in length, which extends from the coast of the Atlantic Ocean as far as that of the South Sea, and which seems to stretch 120 leagues further towards the archipelago of Revillagigedo, covered with tufa composed of fragments of pumice.

These lines of volcanoes, these up-heavings across continued rents, these subterraneous noises (*bramidos y truenos subteraneos de Guanaxuato*, in 1784) which are heard in the midst of a district of schist and transition porphyry, connect, in our imaginations, the still active forces of the New World, with those which in the most remote times heaved up chains of mountains, fractured the surface, and made fountains of liquefied matter (lavas) gush out amidst strata more anciently consolidated. Even in our days this liquefied matter does not constantly issue from the same openings in the orifice of a mountain (crater at the summit of a volcano) or its shattered flank; the earth sometimes (Iceland, table-land of Quito) opens in the plains, from whence currents of lava issue, overflow, cross, and cover each other; or small cones of a muddy substance (moya de Pelileo de Riobamba viejo, February 4. 1797), which seems to have been a trachyte-pumice, and which, being combustible and staining the fingers black, is mixed with the carburet of hydrogen. (Humboldt, *Essai politique sur la Nouv. Espagne*, t. i. p. 47. 254.; Id., *Relat. historique*, t. i. p. 129. 148. 154. 315., t. ii. p. 16. 20. 23.; Klaproth, *Chem. Unterr. der Min.*, t. iv. p. 289.)

The rocks which we are accustomed to arrange together under the name of substances of volcanic formation exclusively, have been hitherto more considered as to the oryctognostic and chemical relations of their composition, or those of their origin, than according to the geognostic connection of their position and their relative age. At every epocha, since the first oxidation of the crust of the globe, the fire of volcanoes has acted across the rocks of the intermediary, secondary, and tertiary formations. With the exception of some freshwater rocks, volcanic rocks alone continue to be formed in modern times. If the lavas of the same volcanoes (the intermitting springs of liquefied matter) vary at different epochas in their eruptions, it may well be conceived that volcanic matter, which during thousands of years has been progressively raised towards the surface of our planet in such different circumstances of mixture, pressure, and cooling, must display both contrasts and analogies. There are trachytes, phonolites, basalts, obsidian and pearlstone of different ages, as there are different formations of granite, gneiss, mica-slate, limestones, grauwacke, syenites, and porphyries.

The nearer we approach to modern times, the more the volcanic formations appear insulated, superadded, and foreign to the soil over which they are spread. A long intermission of the spring seems to produce, even in existing volcanoes, a great variety in the products, and to be opposed to the grouping of analogous substances. In the transition formations (Andes of New Grenada and Peru; Cordilleras of Mexico), the different terms of the geognostic series are connected with each other, and display that mutual dependence which is observed between porphyries and syenites, between clay-slate, greenstone, and transition limestone, and between serpentines, jasper, and euphotide. In this labyrinth of volcanic formations of different ages, few laws of position have been hitherto discovered, which appear, if not general, at least in harmony with the phenomena observed in both continents upon a great extent of country. It is only these relations of position that can here be discussed; all that regards the composition of volcanic rocks, the mechanical analysis of their texture, and their oryctognostic classification, important subjects which have been considered in two celebrated memoirs by M. Fleurian de Bellevue, and M. Cordier (*Journ. de physique*, t. li. lx. et lxxxiii.), does not belong to the domain of the geognosy of formations. Some characters may no doubt be pointed out, in which rocks resemble the productions of modern volcanoes in a more evident manner: but the black colour, the porous and elongated cells covered with a shining coat, the property of forming a jelly with acids, the absence of quartz, common feldspar, and metallic veins (auriferous and argentiferous), the presence of pyroxene, titaneous iron, glassy feldspar, and alcalies, cannot be considered, in the actual state of our knowledge, as the general characters of volcanic rocks. (See above, §§ 21. 23. 26.)

The volcanic masses, or what are considered as such (*empyrodoxian* rocks of M. Mohs, *Charakter der Classen*, 1821, p. 177.), are found either in veins (dykes, in every formation from primitive granite as far as the chalk and tertiary formations; Scotland, Germany, Italy), or in interposed beds (transition limestone and porphyry; red sandstone), or superposed and *superadded* to formations of very different ages. The contrast between intercalated, volcanic, or empyrodoxic rocks, and the rocks in which they are contained, is so much the more striking as the latter are indubitably not volcanic; for instance, calcareous rocks (Derbyshire) or conglomerates (grauwacke, coal-sandstone). When empyrodoxic masses are found, either as subordinate beds, between strata of intermediary crystalline rocks (porphyries and syenites), or as veins traversing the strata of primitive rocks (gneiss-granite), those primitive and intermediary feldspathic rocks may, according to some geognosts, have the same igneous origin as the mass of interposed beds or veins (mandelstein, dolerite, basalt), although the epochas of formation and the circumstances under which the volcanic forces have acted may not have been identical. The distinction between veins and interposed beds of trap, pyroxenic, or porphyritic rocks, are not always so clearly marked as might be supposed from the definitions that are generally given of veins and beds. Several of those beds are only masses formed by the union of a great number of veins. When these follow in great thickness (see my outlines of the celebrated vein of Guanaxuato) the direction and inclination of the strata of any rock, they assume the aspect of a bed. We are the more particular in these remarks, because the new doctrine of geogony has a tendency to lean towards the idea of the liquefied masses ascending across the fissures, from below upwards, while the ancient geogony explained every thing by precipitation, and movements in an opposite direction. We may suppose that these directions must have been different, according to the nature of the consolidated matter, whether crystalline and siliceous, calcareous or fragmentary. Positive geognosy has profited by these discussions on the igneous or neptunian origin of rocks, but it renders the classifications independent of geogonic results; it does not separate the interposed masses of formations in which they are found, and only unites (in the division of rocks which we here consider by the name of *volcanic formations*) the superposed formations, superadded to the primitive, intermediary, secondary, and tertiary rocks.

The place which a rock δ occupies in the geognostic series is determined by *the most recent rock*, γ, which *it covers*, and by the *most ancient* rock, ϵ, *by which it is covered.* If δ be superposed to ϵ, it is quite natural that it should be found placed also on the more ancient rocks α, β, γ, which are the preceding terms of the series. The application of this very simple principle of the geognosy of position requires great circumspection when trachytic, basaltic, and phonolitic rocks are in question. The same current of lava, or the same pyroxenic mass spread over granite, mica-slate, and a fresh-water formation, exhibits no doubt incontestable proofs that its origin is posterior to the most modern tertiary formations; but the age of a volcanic formation is more difficult to determine when there is not a continuity of mass, and when by one general denomination substances that have flowed in a lateral direction, are confounded with others that have come from below, by up-heaving, across pre-existent rocks. Where trachytes and basalts are found united, the latest formation on which the basalts repose does not necessarily determine the age of the trachytes; both these rocks have no doubt been produced in a different manner, and not simultaneously. It is even possible that in a district of small extent, different trachytic masses, which are insulated, but of similar composition, may not be of the same formation, some coming out of transition syenite, others from primitive rocks. The accumulation of trachytic conglomerate most frequently conceals the position of the trachytes so much, that their superposition cannot be ascertained. Thus, the trachytes of

Siebengebirge, near Bonn, are supposed to come out through grauwacke, and those of Auvergne from a table-land of granite, which may belong to the intermediary formation. In the same manner as we should distinguish between real basaltic currents with olivine, and the pyroxenic, spongy, black masses, interposed to the trachytes and to some transition porphyries; so also we must not confound real trachytes (Drachenfels, Chimborazo, Antisana) with feldspathic lavas (leucostiniques) which have flowed in narrow bands (ancient crater of Solfatara of Naples) and which may spread over tufa conglomerates. (Dolomieu, *Journ. des Mines*, Nos. 41, 42. et 69.; Nose, *Niederrhein. Reise*, t. ii. p. 428.; Spallanzani, *Voy. dans les deux Siciles*, t. iii. p. 196.; Ramond, *Nivellem. géogn. de l'Auvergne*, p. 11. 91.; Buch, *Géogn. Beob.*, t. ii. p. 178. 205.; Id., *Mém de l' Acad. de Berlin*, 1812, p. 129—154.; Beudant, *Voy. en Hongrie*, t. iii. p. 508—513. 521—527. et 530—554.)

The trachytic formation in Hungary appears to have been formed between the epocha of the secondary and that of the tertiary formations. M. Beudant, who has written the most complete treatise which we possess on the trachytic rocks, has seen them reposing upon greenstone (Kremnitz, Dregely, Matra) and transition limestone (Glashütte, Neusohl). In Hungary, also, trachytic conglomerates cover grauwacke-slate, and even magnesian limestone, which appears to belong to the Jura formation. In the eastern part of Europe sandstone with lignite, calcaire grossier, and other tertiary rocks, are superposed in their turn on these conglomerates. Similar superpositions of sandstone, gypsum, and limestone of a very recent origin, were observed by M. de Buch and myself at the Canaries and in the Cordilleras of the Andes. The trachytes of the Euganean mountains repose, according to an excellent observer, M. Breislak, on Jura limestone (Schivanoja, near Castelnuovo); but in the country which abounds most in trachytic rocks, the western part of the new continent, both north and south of the equator, I have never seen trachytes pierce through formations so modern.

The most important results of position which my travels in the volcanic zone of the Andes (1801—1804) furnished, may be reduced to the following facts. All the loftiest summits of the Cordilleras are trachytes. The existing volcanoes all act by openings formed in the trachytic rocks. This formation comprehends by zones a great part of the Cordilleras, but rarely extends towards the plains; and the volcanoes which are still burning, far from being solitary or associated by groups of an irregular form, more or less circular, as in Europe (Ramond, *Niv.*, p. 45.; Humb., *Rel. Hist.*, t. ii. p. 16.), follow each other, like the extinguished volcanoes of Auvergne and the burning craters of the Isle of Java, by files, sometimes in one series, sometimes on two parallel lines. These lines are generally placed (mountains of Guatimala, Popayan, Los Pastos, Quito, Peru, and Chili) in the direction of the axis of the Cordilleras; sometimes (Mexico) they make an angle with that axis of 70°. Even where the trachytes do not by their accumulation cover the whole surface, they are found scattered in small masses on the back and the ridge of the Andes, rising in the form of pointed rocks amidst primitive and transition rocks. Trachytes and basalts are rarely united, and those two systems of rocks seem to repel each other mutually. Real basalts, with olivine do not form interposed beds in trachyte, but, where they occur near trachytes (between Quito and the Villa de Ibarra; Julumito at the west of Popayan; valley of Santiago in New Spain; Cerros de las Cuevas and Canoas near the volcano of Jorullo), the latter are covered by basalts and mandelstein. Trachytic rocks are principally seated in the transition formation, in the great formations of syenites and porphyries (§§ 21. and 23.), anterior and posterior to grauwacke and clay-slate, above all, in the first of those formations, which immediately covers the primitive rocks. When, in the Andes, trachytes appear to cover granites, with hornblende, gneiss, and green and steatitic mica-slates, it remains doubtful if these latter rocks, instead of being primitive, do not rather belong to the transition formation. We may consider it as equally problematical whether these appearances of *coverings*, these superpositions of trachytic rocks on pre-existing formations, are not rather simple *appositions*, and if trachyte,

(Extentam tumefecit humum, ceu spiritus oris
Tendere vesicam solet, aut direpta bicornis
Terga capri; tumor ille loci permansit et alti
Collis habet speciem, longoque induruit ævo;
OVID, *Metam.* lib. ix., on the heaved up cone
of Trecene in Argolide,)

in raising up and breaking the ancient crust of the globe, did not issue perpendicularly in the form of bells (Chimborazo), or in that of fortified castles in ruins (summit of the Cordilleras of Peru, between Loxa and Caxamarca). The trachytes of the Andes and of Mexico, which contain pearlstone and obsidian, are generally covered only by other volcanic rocks (phonolites, basalts, mandelstein, conglomerates, pumice-tufas): sometimes small local formations of limestone and gypsum, that may be called tertiary because they are certainly posterior to the chalk, surmount the trachytes; but these trachytes of the Cordilleras towards the lower part, especially if they are not *covered*, are geognostically connected in the closest manner with the cellular transition porphyries, destitute of quartz, and containing pyroxene and glassy feldspar, sometimes rich in argentiferous veins, and in other places supporting secondary formations, even black and carburetted transition limestone (see p. 141. 184. 219. 231.). This connection may be a reason at some future time for suppressing in our systems the *volcanic formation*, inasmuch as it is considered as opposed in the mode of its formation and origin to all other rocks. There are volcanic rocks in the transition series and in the red sandstone, as there are breccias, cemented and united by the action of water, in the volcanic formation. This last term, to give it a precise meaning, should be applied only to the productions of volcanoes, the action of which has been posterior to the existence of our valleys.

Although, according to observations made in both continents, the trachytes and other analogous rocks, which seem also to be owing to the action of volcanic forces, and in which compact or glassy feldspar are more abundant than hornblende and pyroxene, are found *principally* in the transition formation, and on the limits of that formation and the most ancient secondary rocks, we cannot extend this conclusion to basalts, which are often included in primitive granite (Schneekoppe in Silesia; Red Rock, near Serassac in Velay), and which are perhaps anterior to some formations of trachyte. In a very circumscribed country, and in the same grouping of volcanic rocks, the granular trachytes or trachytic porphyries, which we must not confound with breccias or conglomerates of trachytes much more modern, are generally of a far more ancient formation than the basalts which cover them in currents or in vast masses. On the contrary, the basalts posterior to trachytic and pumice conglomerates are most frequently anterior to basaltic conglomerates and tufas; but, we repeat, when we have to compare the scattered outliers of a trachytic, phonolitic, or basaltic formation, outliers not covered, and placed on granitic, intermediary, or secondary formations, those rocks of trachyte, basalt and phonolite, can no longer be ranged as terms of the same geognostic series. What issues from the most ancient granite may be posterior to an analogous rock which has also pierced through transition rocks. Oryctognosy, or descriptive mineralogy, which analyses the texture of volcanic substances, will succeed in classing them according to the principles which M. Cordier has so well established in his memoir *on the composition of pyrogenic rocks of every age;* but geognosy, which considers only the relative age and positions, will be forced to reckon a great number of rocks as *incertæ edis*, even when a more considerable part of the earth shall have been carefully examined. This uncertainty is not owing to the imperfection of the methods of classing, but to the impossibility of comparing, as to their succession or the epocha of their origin, the insulated rocky masses that are *not covered.* The historian of nature, like the historian of the revolutions of the human race, collects, compares, and discusses every fact, but cannot arrange in a series those which present no chronological character.

In this state of things, far from mixing oryctognostic considerations with the classifications of positive geognosy, it appears to me proper to range the volcanic rocks according to the *type of position* most generally observed in both hemispheres, where the greatest number of those rocks are grouped. The great mass of substances in which feldspar predominates (trachytes, leucostines) will be succeeded, as in the oryctognostic tables, by the great mass of substances in which pyroxene predominates (basalts, dolerites); but this apparent agreement between methods founded on two different principles, that of the composition, and that of the order of position, disappears when we examine partial or interposed formations. The geognost then distinguishes between the *phonolites* of *trachytes*, and the phonolites of basalts; he places compact leucostine in the pyroxenic formation, as he arranges a formation of dolerite (a mixture of feldspar and pyroxene, in which the latter substance is the most frequent) amidst leucostine or trachyte. I have sketched, according to these principles, the distribution of volcanic rocks, the table of which is placed at the end of the transition formations (p. 258.). This distribution is founded on the truly geognostic observations published by MM. Leopold de Buch, Breislak, Boué, and Beudant, and on those which I had myself occasion to make in Italy, at the Peak of Teneriffe, and in the Cordilleras of New Grenada, Quito, and Mexico. I shall add to the nomenclature of formations the succinct indication of the most interesting positions of equinoxial America.

I. Trachytic formations, comprehending *granular trachytes* (granitoid and syenitic), *porphyritic trachytes*, or *trachytic porphyries*, partly pyroxenic, partly cellular with siliceous concretions (trachytic meulieres, or *porphyres molaires* of M. Beudant); *semi-vitreous trachytes*, *pearlstone* with obsidian, and *phonolite of trachyte.* To this series may be added the *trachytic* and *pumice conglomerates*, with alumstone, sulphur, opal and opalised wood; for every volcanic formation, like every intermediary and secondary rock, has its conglomerates, that is, its fragmentary rocks, of which it has supplied the first elements. Trachytes (granites heated *in situ*, of the ancient mineralogists, trap-porphyries, many of the petrosiliceous lavas of Dolomieu, domites of MM. de Buch and Ramond, necrolites of M. Brocchi, granular leucostine of M. Cordier) generally present but few traces of stratification in the ancient continent; but in the Cordilleras of the Andes they are often very regularly stratified (Chimborazo, N. 60° E.; Assuay, N. 15° E.), though varying by groups in the direction and inclination, like the phonolites of the basaltic formation (Mittelgebirge in Bohemia). The columnar structure (prisms of 4 to 7 sides) is very common in the porphyritic trachytes of the Cordilleras, not only in the black rocks with a basis of retinite (pechstein) with glassy feldspar and pyroxene (Passuchoa, near the town of Quito, at the south of the hills of Poingasi; Faldas de Pichincha; Paramos de Chulucanas, Aroma, and Cunturcaga, in the Andes of Peru, between Loxa and Caxamarca), but also in the greenish-grey trachytes of Chimborazo (slender prisms fifty feet long; height of the table-land, 2180 toises), as well as in the granitoid trachytes of Pisojè, at the foot of the volcano of Puracè. The latter are greenish-grey, contain black mica, common feldspar, and a little hornblende, and by their resemblance with the *graniti colonnari* of the Euganean Hills, are far removed (p. 167.) from the porphyries of the transition formation. The globular structure (in spheroids with concentric layers) appears to belong rather to the basaltic formations, than to real trachytes. Pale colours predominate in the trachytes of the Cordilleras, and the black masses of that rock appeared to me in general to be posterior to the white, grey, and red masses. The same difference of position seems to occur in Hungary. The black trachytes assume sometimes (Rucu-Pichincha near Quito, principally at the ridge of Tablahuma, 2356 toises) completely the aspect of basalt; but olivine is always wanting, and we only observe small crystals of pyroxene in the interior of the crystals of glassy feldspar. In the Andes, as well as in the ancient continent, every cone or trachytic dome (the former appear like domes or bells, pierced at their summit, and covered on their flanks by pumice and ejections of scoriæ) presents rocks quite different in their composition, according to the element which predominates in the crystalline mass. Black mica is most common in the trachytes of Cotopaxi (between the Nevado of Quelendaña, and the ravine of Suniguaicu, at 2163 toises), a volcano which abounds at the same time in vitreous masses and obsidian; hornblende predominates in the trachytes of Pichincha and Antisana, which are often black; and pyroxene in the lower and middle region of Chimborazo, the trachytes of which sometimes contain pyrites, quartz, and two varieties of feldspar, the glassy and the common. The ancient volcano of Yana-Urcu, close to Chimborazo (on the side of the village of Calpi), is destitute of pyroxene, and contains large crystals of hornblende. In the trachytes of Nevado de Toluca (Mexico) and of Antisana, as in the trachytes of the Puy-de-Dome, spongy and scorified parts with glazed cells are often observed imbedded in compact and earthy masses. The phonolites of trachytes are better characterised in the volcano of Pichincha (Peak of the Ladrillos and Guagua-Pichincha), and also at the eastern declivity of Chimborazo, near Yanacoche (height 2300 toises). At Antisana (Machay of San Simon) and at the north of the Villa de Ibarra (Azufral de Cuesaca, table-land of Quito), trachytes with a base of compact feldspar, mixed with hornblende, contain native sulphur, like the trachyte of the Puy-de-Dome, and the banks of the Dordogne. (Ramond, *Niv. geogn.*, p. 75. 86.) We must not confound this formation of native sulphur with those of solfataras or extinguished craters, cellular mandelstein (between Pate and Tecosautla at Mexico), and clay of the basaltic formation (province de los Pastos). The thickness of these beds of trachytes is such, that, on the table-land of Quito, it attains indubitably, and in *continued masses*, the thickness of 14,000 to 18,000 feet (Chimborazo, Pichincha). Real currents of lithoid lava having issued from very few of the volcanoes of the Andes, the trachytes are there almost every where uncovered. They are only sometimes concealed from the examination of the geognost by trachytic conglomerates and problematic argillaceous formations (tepetate), of which we shall speak shortly.

I found common and milky feldspar in the light, porous, and white trachytes of Cerro de Santa Polonia (1532 toises, near Caxamarca, Andes of Peru), at the summit of Cofre de Perote at Mexico (the Peña del Nauhcampatepetl, 2098 toises), in a reddish-grey trachyte, abounding in acicular crystals of hornblende, and very regularly stratified (N. 28° E. with 30° at N.W.); at the still burning volcano of Tunguragua, at the south of Quito (Cuchilla de Guandisava, 1638 toises), in brick-red and cellular trachytes; and finally at the base of Chimborazo near the small extinguished volcano of Yana-Urcu (1700 toises), in black and vitreous trachytes. M. de Buch, who carefully examined these latter rocks, found in them both crystals of glassy and of common feldspar, a phenomenon which I observed also in several transition porphyries of Mexico.

Small acicular crystals of hornblende are sometimes placed as by files on several parallel lines, and all affect the same direction (valley of Cer at Cantal; whitish-grey trachytes of Riobamba viejo, with rhomboidal feldspar decomposed into a yellowish earth).

Mica is far more rare in the trachytes of Mexico and the Andes, than in those of Siebengebirge, Gleichen in Styria, near Radkersburg, and Hungary); I found, however, fine black hexagonal tables at the base of the volcano of Pichincha (near Javirac or Panecillo of Quito, 1600 toises), as well as in the semi-vitreous bluish-grey trachytes of Cotopaxi, and in the red and porous trachytes of Nevado de Toluca (summit of Fraile, 2372 toises).

Titaneous iron is not wanting in the trachytes of Quito and Mexico; but the tabular crystals of specular iron, common in the trachytes, and also in the lavas of Italy and France, are somewhat rare in the volcanic fissured rocks of equinoxial America.

In considering the trachytes of the Cordilleras under a general point of view, they are no doubt characterised by the absence of quartz in crystals and grains. This character, as we observed above, extends for the most part, even to the metalliferous porphyries of equinoxial America (§§ 23. and 24.), which seem to be connected with trachytes; but both these rocks furnish striking exceptions to a law which might be thought general. These exceptions prove anew that the geognost ought not to annex great importance to the presence or absence of certain substances disseminated in rocks. The greatest mass of the Chimborazo is formed of a semi-vitreous trachyte that is brownish-green (with a base having a waxy appearance like pitchstone), destitute of hornblende, abounding in pyroxene, very compact, tabular, or divided into slender irregular tetrahedral columns. This trachyte contains, as an interposed bed, a purplish-red and cellular bed, with feldspar crystals scarcely visible, and having elongated nodules of white quartz. At a greater height, (3016 toises, where we saw the mercury in the barometer descend to 13 inches, 11 $\frac{2}{10}$ lines), the quartz disappears, and the ridge of the rock on which we trod was covered with a line of red, spongy, disintegrated masses, somewhat similar to the amygdaloids of the valley of Mexico. These masses, taken from the greatest elevation at which specimens have yet been collected on the surface of the earth, were ranged in files, and might lead to the belief of the existence of a small mouth near the summit of Chimborazo, and which is probably closed like that of the Epomeo, at the isle of Ischia, and of Gambalo and Igualata, between Mocha and Penipe (province of Quito).

On the central table-land of Mexico, the trachytes of Lira contain milky quartz, obsidian, and hyalite. M. Beudant has also lately observed crystals of quartz in *porphyritic trachyte* (with vitro-lithoid globules), in the trachytic meuliere, and pearlstone of Hungary. (*Voy. en Hongrie*, t. iii. p. 346. 365. 519. 575.) The same phenomenon is also seen in some trachytes of Auvergne (Puy Baladon, Cantal, Col de Caboe), the Dardanelles, and Kamtschatka. When we recollect that, according to the analysis of M. Vauquelin, there is 92 per cent. of silica in the trachytes of Sarcouy, and that this earth abounds in basalt and lava, we may rather be surprised that this substance, disseminated in the silicates of iron and alumine, has not united together oftener without mixture into crystals or grains of pure quartz. This difficulty which has been opposed to the concentration of silica around a centre characterises a great part of the volcanic rocks. (See above, p. 160.)

Pyroxene has hitherto been regarded as extremely rare in the trachytes of Europe. The bed of pyroxene discovered by M. Weiss, between Muret and Thiezac (above Aurillac in Auvergne; Buch, *über Trapp-Porphyr*, p. 135.), seems rather to belong to a basaltic formation superposed to trachyte. But in Hungary (Beudant, t. iii. p. 317. 519.), as well as in the Cordilleras of the Andes, pyroxene is often found in porphyroid trachytes, where it replaces hornblende (Chimborazo, Tunguragua, base of the volcano of Pasto, middle region of the volcano of Purace, near Popayan). The kind of repulsion which is observed between pyroxene and hornblende, is so much the more striking, as in the basaltic formation these two substances are often found united (Rhönegebirge in Germany). The trachytes of Mexico appeared to me in general to be destitute of pyroxene.

Garnet, which we have already seen in the transition porphyries of Potosi and Izmiquilpan, appears also, although very rarely, in the trachytes of the Andes. I found it in the volcano of Yana-Urcu (black trachyte), and M. Beudant collected it in the lithoid pearlstone of Hungary.

I now doubt the existence of olivine in the trachyte formation of the Cordilleras; what I had taken for that substance were grains of pyroxene of a light tint. Olivine belongs perhaps exclusively to the basaltic formations, and some lithoid lavas. M. de Buch recognised it among the ejections of the volcano of Jorullo, which formed a fine-grained mixture of olivine, glassy feldspar, and yellow mica: there is no trace of hornblende and pyroxene, although this volcano has pierced through a trachytic formation. M. Beudant also doubts the presence of olivine in the trachytes of Hungary, even in those of the group of Vihorlet. When chemists shall examine with more care the trachytes of the Cordilleras, which present so great a variety of rocks, muriatic acid will probably be discovered (as at Sarcouy in Auvergne), and common mica mixed with oxyde of titanium, as at Vesuvius. (Soret, *Sur les axes de double refraction*, 1821, p. 59.)

The observations which may be made on the position of volcanic rocks promise more interest than even the study of their composition. The trachytes of the extinguished volcano of Tolima (§ 7.) seem to issue from a granite posterior to primitive gneiss. I saw mica-slate (Alto del Roble) appearing (p. 108.) below the trachytes of the still burning volcanoes of Popayan. The granites, through which the trachytic domes of Baraguan and Herveo (Ervè) have pierced are perhaps of a more recent date than mica-slate. The most important observation of position which I made in the immense and entirely trachytic table-land of Quito (species of polystome volcano) relates to the trachytes of Tunguragua. After having sought in vain, during more than six months, some trace of rocks commonly supposed of neptunian origin, I found, near the bridge of ropes of Penipe (Rio Puela, 1240 toises), below a black semi-vitreous trachyte, often columnar, at the still burning cone of Tunguragua, a greenish mica-slate, with a striated and silky surface, containing garnets, and resembling the mica-slate of the primitive formation. (See above, p. 93.) This rock reposes on syenitic granite composed of a great quantity of greenish lamellar feldspar in large grains, a little white quartz, hexagonal tables of black mica and some thin crystals of hornblende. The fracture of this granite has a steatitic aspect, and, when breathed upon, takes a tint of asparagus-green. These syenites and mica-slates with garnets remind us of those which MM. de Buch and Escolar discovered in the archipelago of the Canaries, in blocks amidst the trachytic formations of Fortaventura and Palma. (Humboldt, *Rel. Hist.*, t. i. p. 640.) It is very certain, that the rocks of Penipe, which perhaps belong only to the transition formation, are *in situ*; that they appear beneath a real granular trachyte, and not beneath a breccia or trachytic conglomerate, as at Vic, Aurillac, and S. Sigismond (Buch, *Trapp-Porphyr*, p. 141.); but it is impossible to decide on the superposition without piercing a gallery in the flank of Tunguragua, whether the trachyte covers mica-slate on a great extent, as the chalk covers the Jura limestone, or if the trachyte in breaking through more ancient rocks and rising perpendicularly, is simply inclined towards the border on the adjacent mica-slate. We also find around the trachytic cone of Cayambe, mica-slate with epidote, and a granite that abounds in brown and yellow mica. Farther north, in the Cordilleras of Popayan, ascending to the village of Puracè, I saw beneath a great volcano of that name, near Santa-Barbara, semi-vitreous trachyte leaning on porphyritic syenite (with common feldspar); this syenite is visibly superposed on a transition granite abounding in mica (p. 178.). At the foot of the still active Mexican volcanoes of Popocatepetl and Jorullo, M. Bonpland and myself were not fortunate enough to discover rocks of granite, mica-slate, and syenite on the spot; but we saw imbedded amidst the lithoid, black, and basaltic lavas of Jorullo, white angular or greenish-white fragments of syenite, composed of a little hornblende, and a great deal of lamellar feldspar. Where these masses have been split by the heat, the feldspar has become filamentous, so that the edges of the fissures are in some places joined by the lengthened fibres of the mass. In South America, between Almaguer and Popayan, at the foot of Cerro Broncaso, I found real fragments of compact gneiss in a trachyte abounding in pyroxene (p. 171.). These phenomena, to which I might add many others, prove that trachytic formations have issued from below the granitic crust of the globe.

The obsidian, of which M. Sonneschmidt and myself brought such curious varieties into Europe, appeared to me to belong, in the Cordilleras, to two very distinct sections of trachytic formation, to real white (Cerro de las Novajas or Oyamel, at the north-east of Mexico), to black trachytes (Cerro del Quinche, north of Quito), and to pearlstone (Cinapecuaro, between Mexico and Valladolid). We must distinguish in those two formations, the obsidian from the currents of modern lava (Peak of Teneriffe) forming the upper part of those currents. The fragments of rock ejected by the crater of Cotopaxi, and filled with nodules of obsidian, appear to have been torn from the sides of the crater; but the pieces of obsidian thrown out by the volcano of Sotara near Popayan, to the distance of several leagues, deserve more attention; they cover the fields of Los Serillos, Uvales and Palacè. They are found scattered like fragments of flints; they lie on basaltic rocks, to which, however, they are altogether foreign. These obsidians of Popayan have often the form of tears, or of balls with a rough surface; they exhibit, which I have not seen any where else, every shade of colour, from a deep black to that of colourless glass. They are sometimes mixed with fragments of enamel, thrown out also by the volcano of Sotara, and which one might be inclined to take for *Reaumur's porcelain.* The paste of the semi-vitreous bluish-grey trachyte, with a conchoidal fracture (volcano of Puracè, near Popayan, in the plain of Cascajal, at 2274 toises high), sometimes, no doubt, passes into obsidian; but the great masses of real obsidian, disposed by beds or by nodules, with outlines strongly marked, are found in other varieties of trachytes. We have already described above, the rocks of Cerro de las Navajas (§ 23.), where obsidians are found that are chatoyant, striated, and silvery (*plateadas*), and generally scattered in fragments, but sometimes, also, forming beds in a white trachyte. Similar beds, from fourteen to sixteen inches of thickness, are interposed in the black pyroxenic trachytes of Cerro del Quinchè (table-land of Quito). They afford greenish-black obsidians, veined with bands of brick-colour. Near Hacienda de Lira, at the north of Queretaro (table-land of Mexico, 995 toises), I found in an olive-green trachyte, with a base of retinite (trachytes which contain both glassy feldspar and grains of disseminated quartz), beds of black obsidian three inches thick. In other parts of the table-land of New-Spain, at Cinapecuaro, at the foot of Cerro Ucareo (in the way from Valladolid de Mechoacan to Toluca, 968 toises high), and between Ojo del agua and El Pinal (in the road from la Puebla de los Angeles to Perote, 1180 toises high), obsidians are found as nodules in a pearlstone, with a lustre like enamel, composed of small globules that are semi-vitreous and greyish-white. I saw no mica, but infiltrations of hyalite, and some small crystals of feldspar that are filamentous and almost of the structure of pumice. Pearlstone forms, at Cinapecuaro, little conical hills surrounded with peaks of basalt and trachytic domes. The rock is very regularly stratified (N. 22° E., incl. 80° at north-west), and might be taken at a distance for a schistose sandstone. Black, darkish-green, and greyish-green obsidian occurs in nodules, from two to five inches thick; so that by the juxta-position of those nodules, the pearlstone appears sometimes to be imbedded in a real obsidian rock. In the eastern plains of Mexico, between Acaxete Ojo del agua and El Pinal, obsidian is less abundant, but often striped like jasper. Pearlstone contains a great quantity of hexagonal tables of black mica; it is often fibrous, and passes to what M. Beudant calls (t. iii. p. 364. 389.) *perlite ponceuse.*

The obsidians of Mexico and the Andes of Quito exhibit, in general, and often on a greater scale, the same phenomena of composition which is observed in those of Lipari and Volcano; and which some geognosts formerly attributed to a *devitrification* (*glastinisation*). We there find imbedded, small crystals of glassy feldspar; polyhedral masses of pearlstone filling up entirely the cavity in which they are supposed to be formed; aggregations of dark grains, of an earthy aspect, and arranged in parallel zones, not continuous; finally, fragments of reddish-brown trachyte, half melted, all placed on the same side, at the extremity of cavities that are very much elongated, and parallel to each other. M. de Buch, who has particularly examined the volcanic substances which I collected in the equinoxial region of the New World, observes, that masses of pearlstone, sometimes spheroidal, sometimes octagonal in their section, have constantly at their centre a very small crystal of glassy feldspar, or hornblende, and that the position of that crystal has determined the form of the whole system. (Buch, in the *Schriften Naturf. Freunde*, 1809, p. 301.; Humboldt, *Rel. Hist.*, t. i. p.161.) M. Beudant found red garnets in the retinitic pearlstones of Hungary (Vissegrad), which resemble the *pitchstone-porphyry* of the transition formation; I also saw red garnets at the summit of the volcano of Puracè, in a bluish, semi-vitreous trachyte, with a conchoidal fracture, destitute of mica and hornblende, but containing, besides pyroxene and glassy feldspar, ash-coloured points, similar to those that are remarked in the obsidians of Lipari, and of Cerro de las Navajas. The presence of garnets in rocks generally mixed with hornblende, derives some importance from the ingenious observations of M. Berzelius (*Nouv. Système de Minéralogie*, p. 301.) on the chemical affinities of garnet and hornblende, containing silicates of alumine and oxidulated iron. M. Collet-Descotils found in the obsidians which I brought from New Spain, the first example of the simultaneous presence of two alkalies in the same mineral substance. This phenomenon has since been observed in some varieties of feldspar, wernerite, sodalite, chabasie, and eleolite (pierre grasse of Haüy). I observed that many black and red obsidians of Quinchè and Cerro de las Navajas, have magnetic poles; like the porphyries (of transition? p. 176.) of Voisaco, and like a fine group of columnar trachytes of Chimborazo (2100 toises high). These trachytes were greenish-grey, and contained some crystals of lamellar and milky feldspar.

The last division of the trachytic formation is formed by conglomerates, or agglutinated remains, re-formed by the waters. These conglomerates cover immense surfaces, not at the foot of the Cordilleras, but on their flanks, and on table-lands from 1200 to 1600 toises high. In a region where almost all the burning volcanoes rise above the limit of perpetual snow, and where the waters slowly filtered in caverns, and the snows which dissolve at the moment of the eruption, cause dreadful ravages, the extent and thickness of alluvial formations, and regenerated fragmentary rocks, must necessarily be in proportion to the forces which still move those disaggregated masses in modern times. The conglomerates are sometimes friable like tuffa (base of Cotopaxi and the l'Altar), sometimes compact and hardened like sandstone (base of Pichincha). The pumice in pulverulent masses, and in blocks from twenty-five to thirty feet long, forms the most interesting part of these conglomerates of trachytic formation. We shall observe on this occasion, that the word *pumice* is very vague in mineralogy; it does not designate a simple mineral like the denominations of calcedony or pyroxene; it rather denotes a certain state, a capillary or fibrous form, under which the various substances ejected by volcanoes appear. The nature of those substances is as different as the thickness, the tenacity, the flexibility, and the parallelism, or the direction of their fibres. (Humboldt, *Relat. Hist.*, t. i. p. 162.) There is a black pumice of a spongy texture, with interwoven fibres, where we see a great deal of pyroxene, and which seem to owe their origin to basaltic scorified lavas (plain that surrounds the crater of Rucu-Pichincha; tuffa of Pausilippe near Naples). Some volcanoes eject white trachytes, composed of compact feldspar, much hornblende, and very little mica, of which a part is fibrous (Rucu-Pichincha and Cotopaxi, on the table-land of Quito; volcano of Cumbal near Chilanquer, in the table-land of Los Pastos; Sotara near Popayan; Popocatepetl at the east of Mexico). In trachytes somewhat compact, and of a texture not fibrous, the rhomboidal fragments of feldspar often become hollow and fibrous (table-land of Quito and Mexico). Some varieties of pearlstone present a fibrous texture (plain of New Spain, between la Ventu del Ojo del agua and la Ventu de Soto; valley of Gran and Glashütte in Hungary); finally, obsidians of a greenish-black or smoke-grey, alternating with beds of pumice, with asbestoid greenish-white fibres, rarely parallel with each other, sometimes, however, perpendicular to the strata of obsidian, and similar to a filamentous froth of glass (Plain of Genets at the Peak of Teneriffe). These latter varieties led some geologists to think that all pumice was derived from the fusion and bubbling up of vitreous lava; they confounded the pumice-obsidians (asclerines of M. Cordier) with the true pumice with parallel fibres (light pumice of M. Cordier), characterised by large hexagonal plates of mica, and probably owing to a particular mode of action which the fire of volcanoes exerts on white trachytes (granites of the Isles Ponces of Dolomieu). A geognost who has studied profoundly the trachytic rocks of Europe has confirmed this view. "Pumice," says M. Beudant, "in the present state of the science, cannot even be regarded as a distinct species of rock; it is a cellular and filamentous state which several rocks of trachytic and volcanic formation are capable of assuming." (*Voyage Minéral.*, t. iii. p. 389.)

The immense subterraneous quarries of pumice worked at the foot of Cotopaxi, between the town of Tacunga (Llactacunga) and the Indian village of San Felipe (table-land of Quito, 1482 toises high), appeared to me the most instructive for deciding on the position of that substance in an alluvial soil. They had already given rise in the mind of Bouguer (*Figure de la Terre*, p. lxviii.), at a time when geognosy scarcely existed, to several interesting questions on the origin of pumice. The little hills of Guapulo and Zumbalica, which rise to the height of 80 toises, appear at first sight entirely formed of a white fibrous rock, with horizontal beds and perpendicular fibres; from which blocks without rents might be taken, more that sixty feet long. In examining more closely these pretended beds, they are perceived to be masses from four inches to three feet thick, imbedded in a white clayey earth. They do not form a conglomerate, properly speaking; the blocks are only deposited in clay, and covered with small fragments of pumice (from eight to nine inches thick), which are divided into horizontal beds. These blocks of white pumice, which are sometimes bluish, are rounded towards the edges; they contain yellow and black mica, slender crystals of hornblende (not pyroxene), and a little glassy feldspar. I am inclined to think that the hills of Zumbalica, which much resemble those of Sirok in Hungary (Beudant, *Voy. Minér.*, t. ii. p. 22.), are not the interior walls of an overthrown volcano; the great blocks which resemble fractured beds are geognostically connected with the small fragments of the upper strata; both have, no doubt, been deposited by water, although under circumstances very different from those which accompany the actual eruptions of Cotopaxi. The aspect of the whole surrounding country proves the ancient sphere of activity of this volcano, which is 2952 toises high, and of an enormous volume. At the west of the volcano, from l'Alto de Chisinche as far as Tacunga, for more than forty square leagues, the whole soil is covered with pumice and scorified trachytes.

It is very remarkable, that the mode of volcanic action proper for the production of pumice is restricted in some degree to a certain number of ignivome mountains. L'Altar or Capac-Urcu, anciently more lofty than the Chimborazo, is placed in the plain of Tapia, opposite to the still burning volcano of Tunguragua. The former has ejected an immense quantity of pumice, the latter produces none. The same difference exists between the two neighbouring volcanoes of the town of Popayan, Puracè and Sotara; which latter has ejected at the same time obsidian and pumice, like the volcano of Cotopaxi. At Rucu-Pichincha, where I reached one of the trachytic towers (2491 toises high) which commands the crater of the volcano, I found a great quantity of pumice, and no obsidian; the pumice, therefore, of Sotara and Cotopaxi, which contains, besides glassy feldspar a little amphibole, and large hexagonal plates of mica, is certainly not owing to obsidian; it differs entirely from the vitreous and capillary pumice which I saw covering the declivity of the Peak of Teneriffe.

The fine opals of Zimapan, at Mexico, do not appear to belong, like those of Hungary, to trachytic conglomerates, but to porphyritic trachytes which contain radiating globules of bluish-grey pearlstone. (§ 23.)

II. Basaltic formations, comprehending *basalts* with olivine, pyroxene, and a little hornblende; *phonolite of basalt, dolerite, cellular amygdaloid, clay with pyrope-garnets,* and *fragmentary basaltic rocks* (conglomerates and scoria). The basaltic formation is connected on one side with trachytes, in which pyroxene becomes progressively more abundant than feldspar (Cordier, *sur les Masses des Roches Volcaniques,* p. 25.), and partly, and I believe in a closer manner, to the lavas of volcanoes which have flowed in the form of *currents.* Phonolites belong both to trachytic and basaltic formations; I doubt if a real basalt with olivine is found interposed as a subordinate bed to trachyte. The phonolite which forms these beds in the trachytes of the Cordilleras and of Auvergne, is only superposed to basalt. When it does not rise in insulated peaks in the plains, it generally crowns the basaltic hills. Hornblende and pyroxene are found disseminated in trachytes and basalts; the former of those substances belongs perhaps even more particularly to the trachytic formations. Olivine characterises basaltic rocks, the very ancient lavas of Europe, and the very modern lavas (in 1759) of the volcano of Jorullo, at Mexico.

When we consider the group of trachytic and basaltic rocks only in relation to their size, we observe that the great masses of those groups are found very remote from each other. The countries that abound most in basalts (Bohemia and Hesse) have no trachytes, and the Cordilleras of the Andes, which are trachytic for an immense extent, are often entirely destitute of basalts. Neither Chimborazo, nor Cotopaxi, Antisana, nor Pichincha, furnish any real basaltic rocks; while those rocks characterised by olivine, divided into fine columns three feet thick, are found on the same table-land of Quito, but far from those volcanoes at the east of Guallabamba, in the valley of Rio Pisque. The basalts near Popayan do not cover the trachytic domes of Sotara and Puracè; they are found insulated on the western bank of Cauca, in the plains of Julumito. The great basaltic formation of the Valle de Santiago in Mexico (between Valladolid and Guanaxuato), is very distant from the trachytic volcanoes of Popocatepetl and Orizava. All the basalts we have just named (Guallabamba, Julumito and Santiago) probably repose also at great depths on trachytic rocks; but we shall not consider here particularly the separation of *mountains* of basalt and of trachytes.

In the Cordilleras of Mexico, New Grenada, Quito and Peru, the mass of trachytic formations is in general far more considerable than that of the basaltic formations; the latter may even be considered as very rare, when compared with those which traverse Germany from east to west, between the parallels of 50° and 51°. The same preponderance of trachytic formations over basaltic rocks, is observed in Hungary. "Wherever," says M. Beudant, with great justice, "masses of trachyte are developed on a great scale, we find only inconsiderable fragments of basalt, and reciprocally, in the places where the basaltic formation is much developed, there occurs little or no trachyte. (*Voyage Minér. en Hongrie*, t. iii. p. 500. 587—589.) It would seem as if those two formations repelled each other, and as the craters of existing volcanoes are constantly open in trachytes, we must not be surprised if those volcanoes and their lavas remain so far removed from the ancient basalts. (Humboldt, *Rel. Histor.*, t. i. p. 154.)

Notwithstanding this opposition, or rather this inequality of developement, which we have already remarked in granite and mica-slate-gneiss, in limestone and transition slate, in red sandstone and zechstein or alpine limestone, trachytes and basalts display in some parts of the globe the closest geognostic affinities. If the great basaltic masses (Hesse, Forez, Velay and Vivarais; Scotland; Vezprim and the lake Balaton) remain geographically removed from the great masses of trachytes (Siebengebirge; Auvergne; Mountains of Matra, Vihorlet, and Tokay; western Cordillera of the Andes of Quito), outlines of the basaltic formation are not the less, on that account, superposed to those very trachytes. (Buch, *Briefe aus Auvergne*, p.289.; Id., *Trapp-Porphyr*, p.137—141.; Ramond, *Niv. Géologique*, p. 18. 60—73.) The Euganean Hills (basalts of Monte Venda, near the trachytic cones of Monte Pradio, Mont Ortone and Monte Rosso), the slopes of the mountains which constitute the group of Mont Dore, the vicinity of Guchilaque in Mexico (Cerro del Marquès, 1537 toises), and of Xalapa (Cerro de Macultepec, 788 toises), furnish striking examples of the union of feldspathic and pyroxenic formations. Sometimes they are conical hills of prismatic basalt, that issue from a trachytic formation; sometimes they are large currents of basalt, often interrupted and forming steps, which furrow and spread over the formation.

It results from these observations, that the greatest masses of basalt are placed immediately in primitive, intermediary, and secondary formations; while other masses, much less considerable, and of a texture altogether identical, and exhibiting most frequently the appearance of ancient currents of lithoid lavas, are superposed on trachytic rocks. Both of these sometimes envelope fragments of granite, gneiss, or syenite abounding in feldspar. The same phenomenon, as we have seen above, is observed (volcano of Jorullo) in recent lavas of a known epocha; but those incontestable indications of an igneous fluidity do not warrant us in admitting that the conical mountains of basalt dispersed in the plains, or crowding the ridge of primitive mountains, are all formed like the masses of basalt that cover trachytes, or like the lithoid basaltic lavas (with olivine) of some very modern volcanoes. The mixture of substances that constitute the volcanic rocks is found in the interior of the globe, and probably at immense depths. Substances analogous and composed of the same elements may appear (at the surface of the globe) by very different means; sometimes by heaving up (in domes or conical elevations); sometimes by longitudinal fissures formed in the crust of the globe; sometimes by circular openings at the summit of the mountain. The geognosy of volcanoes distinguishes all these modes of formation, and if it does not admit all the rocks of trachytic and basaltic formation under the name of *lavas*, it is because it does not admit that the domes of Puy de Cliersou, the great Sarcouy, and the Chimborazo, as well as the conical mountains of basalt, are portions of currents of lava. Some volcanoes, partly very modern, have thrown out feldspathic and pyroxenic lavas (Ischia, Solfatare de Pouzzole) with olivine (Jorullo), which have the greatest resemblance to the most ancient trachytes and basalts. Very often volcanic masses, when considered mineralogically (feldspathic and pyroxenic lavas, trachytes, basalts in insulated cones), are the same; it may be supposed that the circumstances in which they were produced in the interior of the globe differed very little; but what removes them geognostically from each other, is the marked difference in the mode of their appearance at the surface of the earth.

Among the great number of curious observations which the vicinity of the new volcano of Jorullo in Mexico affords, none appeared to me more important and more unexpected than those which relate to the double origin of basaltic masses.

We there see at the same time small cones of basalt, composed of balls with concentric layers, and a promontory of basaltic lava, lithoid and compact within, and cellular on the surface, This current of lava is a black mass with very small grains, not containing hornblende or pyroxene, but certainly olivine (péridote granuliiforme, Haüy) and small crystals of glassy feldspar. M. de Buch found in the fragments which I brought home, besides disseminated olivine (light olive-green, conchoidal, and in separate granular pieces), some hexagonal plates of yellow mica. In those lavas the angular and fissured fragments of granitic syenite, which I have several times mentioned, are imbedded: they probably derive their origin from a transition formation placed beneath trachyte. Very small pieces of greyish trachyte, with glassy feldspar and slender crystals of hornblende, which we were so fortunate as to find on the brink of the crater, in the midst of scoriæ, prove that the eruption has acted at the same time across the syenite and the superposed trachyte. The lavas are even 678 feet in thickness, and as they are not spread laterally, but come from the crater of the burning volcano, it was in following their course towards the S.S.E. that M. Bonpland and myself penetrated, not without some danger, into the interior of the crater to collect air. We must not confound with this current of basaltic lithoid lava, which is not a heap of scoria as at Monte Nova near Pouzzole, the basalts in balls (Kugelbasalt) which compose the small cones called *ovens* (hornitos) by the natives, on account of their form, and because they disengage from fissures thin streams of aqueous vapours mixed with sulphureous acid. No doubt can remain, even with an observer little accustomed to the aspect of rocks disturbed by the fire of volcanoes, that the whole soil of *Mal-pais*, which is at least 1,800,000 square toises, has been thrown up. Where this ejected matter is contiguous to the plain of the *Playas of Jorullo*, which has undergone no change, and of which formerly it made a part, there is (at the east of San Isidoro) an abrupt descent from a perpendicular height of twenty-five to thirty feet. The darkish and argillaceous beds of *Mal-pais* appear as if fractured, and display, in a curved directed from N. E. to S. W., the fissures of stratification horizontal and undulated. After having passed this stage, you ascend on a surface swelled in the form of a bladder, towards the crevice from which the great volcanoes issued, and of which one alone, that of the middle (*El Volcan Grande de Jorullo*), is still burning. The convexity of this formation is in some places 78, in others 90 toises; that is, the foot of the Great Volcano, or rather the central portion of the plain of *Mal-pais* whence the Great Volcano rises abruptly (near the ancient Hacienda of San Pedro de Jorullo), is nearly 510 feet more elevated than the bank of *Mal-pais* near the first stage. The whole declivity of this convex surface is so gentle, that it may escape the attention of those who are not provided with instruments proper for its measurement. It is, as the natives well say, a *hollow soil*, a *tierra hucea.* This opinion is confirmed by the sound of the horses' feet when they they pass over it, by the frequency of fissures, by partial sinkings, and by the engulfed rivers of Cuitimba and San Pedro, which are lost at the east of the volcano, and reappear, as thermal waters of 52° cent., on the western bank of *Mal-pais.* The beds of black or yellowish-brown clay have been heaved up; the surface of the soil is covered only with some volcanic ashes, no accumulations of scoriæ or ejections issuing from the crater have caused the convexity of *Mal-pais.* From this heaved up soil (Sept. 1759) several thousands of small cones or basaltic hills with very convex summits have issued (the *ovens*, or *hornitos*). They are all insulated and scattered, so that in order to approach the foot of the great volcano, you pass by winding paths (*los callejones del Mal-pais*). Their elevation is from six to nine feet. The smoke issues a little below the point of the cone, and remains visible for fifty feet in height. Other streams of smoke come from large crevices which cross the paths; they belong to the soil itself of the heaved up plain. In 1780, the heat of the *hornitos* was still so great that a cigar could be lighted by tying it to a stick and plunging it two or three inches deep in one of the lateral openings. The cones (*hornitos*) are uniformly composed of spheroidal basalt, often flattened, from eight inches to three feet of diameter, and imbedded in a mass of clay in contorted beds. The aspect of these cones is absolutely the same with that of the conical hills of globular basalt (*Kugelbasalt-Kuppen*) which are so frequently seen in Saxony, on the frontiers of the Upper-Palatinate and Franconia, and principally in the Mittelberg of Bohemia; the difference consists only in the dimensions of the hills. M. Freiesleben and myself saw some, however, that were perfectly insulated, and were only from fifteen to twenty feet high. The centre of the balls in the hornitos, as in the ancient globular basalts, is somewhat fresher and more compact than the concentric strata which envelope the nucleus, and of which I was often able to count from twenty-five to twenty-eight. The whole mass of these basalts, constantly traversed by acidulated and hot vapours, is extremely decomposed. It often presents only a black and ferruginous clay, with yellow spots, perhaps too large to be attributed to a decomposition of olivine. In approaching the ear to one of those cones, a hoarse sound is heard, like that of a subterraneous cascade; it is, perhaps, caused by the waters of the Rio Cuitamba, which are engulfed in the *Mal-pais*. Here, therefore, are certainly flattened spheroids of basalt, agglomerated into conical hills, which have been heaved up in our times, and which, consequently, are neither fragments of ancient currents of lava, nor the result of a decomposition of basaltic articulated prisms, nor that of a fortuitous accumulation of ejections of a distant crater. It is probable that the elastic force of vapours has covered the convex plain of *Mal-pais* with these *hornitos*, in the form of blisters, just as the surface of a viscous fluid is covered with bubbles by the action of gas which disengages from it. The crust that forms the small domes of the *hornitos* has so little solidity that it sinks beneath the fore-feet of a mule in ascending.

The facts which I have just stated appear to me so much the more important to geognosy, as there exists in the most ancient basaltic formations a great analogy between the insulated hills of globular basalt and those of columnar basalt. Some celebrated geologists have long combated the hypothesis which considers so many basaltic mountains of so regular a form, and symmetrically grouped, as the remains of a current, in flowing of lava, which has advanced progressively on a sloping ground. We must distinguish three great phenomena in the plains of Jorullo; the general heaving up of the *Mal-pais*, covered by several thousands of small basaltic cones; the accumulations of scoriæ and other incoherent substances in the hills the most remote from the great volcano, and the lithoid lavas which the mountain has thrown out in the ordinary form of a current. The interior of the crater of Vesuvius, which I visited several times in the month of August 1805, with MM. de Buch and Gay-Lussac, exhibited the same difference between the bottom of the heaved up crater, that is to say, more or less convex, as we approach the epocha of the great eruption, and the cones of aggregated scoriæ which are formed around several burning openings. It is these accumulations of incoherent matter alone which resemble Monte novo of Pouzzole. The crust of the lava that constitutes the bottom of craters rises and sinks like a moveable floor (Buch, *Geogn. Beob.* t. ii. p. 124.) At Vesuvius the bottom was so rounded (in 1805) that its central part overpassed the level of the southern edge of the volcano. The swelling which is periodically observed in the accessible waters of burning volcanoes, at the bottom of a circular and lengthened valley that terminates their summits, furnishes a striking analogy with the *heaved up soil of the Mal-pais* of Jorullo; and probably also, with those volcanic islands which appear like *black rocks* above the surface of the ocean, before they burst and throw out flames. It appears that M. d'Aubuisson had no opportunity of consulting the sections which I published of the volcano of Jorullo (Humboldt, *Essai politique*, t. i. No. 253.; Id., *Nivellement barom. des Andes.* No. 370—374.; Id., *Vues des Cordillères*, p. 242. pl. 43.; Id., *Atlas géographique et physique du Voyage aux rég. équin.* pl. 28. et 29.), when, in his interesting *Traité de géognosie*, t. i. p. 264., he supposes that I confounded a heaved up soil with accumulations of ejections, of which the thickness increases as we approach the volcanic mouth.

The composition of basalt, or rather the more or less frequent occurrence of some crystallised substances disseminated in it, varies in different parts of equinoxial America, as well as in Europe. Olivine, so common in the basalts of Germany, France, and Italy, is very rare, according to MM. Macculloch and Boué, in the west of Scotland, and the north of Ireland. Hornblende, in large crystals, abounds in Saxony (Oberwiesenthal and Carlsfeld), in Bohemia, in Fulde, and in Hungary (Medwe), while it is most frequently wanting in the basalts of Auvergne and the Canaries. Glassy feldspar and olivine are found almost constantly associated in the basaltic formation of Mexico and New Grenada; often (Valle de Santiago, Alberca de Palangeo) hornblende and pyroxene are wanting; at other times (Cerro del Marquès, above San Augustin de las Cuevas; Chichimequillo near Silao) basalt contains olivine, glassy feldspar, amphibole, and pyroxene. In the fine valley of Santiago (New Spain) hyalite is so common that, by a predilection very difficult to explain, the ants collect it whenever the basalt is decomposed, and transport it to their nests. I never saw very great masses of olivine in the Cordilleras of the Andes; those of Europe belong more particularly to basaltic breccia (Weissenstein near Cassel; Kapfenstein in Styria).

The formations of clay and marl, which we have mentioned in the preceding table as belonging to volcanic formations, merit great attention in the Cordilleras of the Andes, the archipelago of the Canary Islands, and in the Mittelgebirge of Bohemia (Trzeblitz, Huvka). In those three regions, which I visited successively, the clay did not appear to me to be accidentally placed in the liquid mass, as is sometimes the case in the plastic clay (sandstone with lignites, § 35.) above the chalk, or in the secondary and tertiary limestone (Jura and coarse limestone) of the Vicentino, which I found imbedded as angular fragments in basalt, and which penetrate the basalts so much, that even the latter effervesce with acids. The argillaceous marl of the Cordilleras (Cascade of Regla, and road from Regla to Totomilco el grande; Guchilaque, at the north of Cuernavaca; Cubilete, near Guanaxuato), and those of the Isle of Graciosa (near Lancerote), alternate with beds of basalt, and are perhaps of a contemporary formation, like the argillaceous slates that alternate with alpine limestone. (Humboldt, *Relat. Hist.*, t. i. p. 88.) Their position even seems to prove, that they are not owing to the decomposition of basalt; crystals of pyroxene and pyrope-garnets are often found in them. I shall not decide whether the masses of clay, which in the Andes of New Grenada surround (between Popayan, Quilichao, and Almaguer) those immense heaps of balls of dolerite and greenstone with glassy feldspar, belong to formations of basalt or to syenite and porphyry of the transition formation; but undoubtedly the beds of clay (*tepetate*), which render a part of the fine province of Quito sterile, have issued from the flanks of volcanoes, not mixed with matter in fusion, but suspended in water. The inundations that always accompany the eruptions of Cotopaxi, Tunguragua, and other volcanoes of the Andes still burning, are not owing, as at Vesuvius (*Mémoires de l'Académie*, 1754, p. 18.), to torrents of pluvial waters poured from the clouds that form during the eruption (by the disengaged vapour of water in the crater); they are principally the result of the melting of the snows, and the slow infiltrations that take place on the declivity of volcanoes, of which the height surpasses 2460 toises (that of the limit of perpetual snow). The shocks of violent earthquakes, which are not always followed by flames, open caverns filled with water, and those waters bear along with them broken and bruised trachyte, clay, pumice, and other incoherent substances. These perhaps might be called *muddy eruptions*, if that denomination did not connect a phenomenon of inundation too nearly with phenomena essentially volcanic. When (on the 19th June, 1698) the Peak of Carguairazo sunk down, more than four square leagues around were covered with *clayey mud*, called in the country *lodazales*. Small fish, known by the name of *Prenadillas* (*Pimelodes cyclopum*), a species which inhabits the streams of the province of Quito, were enveloped in the liquid ejections of Carguairazo. These are the fish said to be thrown out by the volcano, because they live by thousands in subterraneous lakes, and, at the moment of great eruptions, issue through crevices, and are carried down by the impulsion of the muddy water that descends on the declivity of the mountains. The almost extinguished volcano of Imbaburu ejected, in 1691, so great a quantity of *Prenadillas*, that the putrid fevers which prevailed at that period were attributed to miasma exhaled by the fish. (Humboldt, *Recueil d'obs. de zoologie et d'anatomie comparée*, t. i. p. 22. and t. ii. p. 150.)

Dolerite of the basaltic formation (D'Aubuisson, *Journ. des mines*, t. xviii. p. 197.; Leonhard and Gmelin, *vom Dolerit*, p. 17—35.) is very rare in the Cordilleras, which rather abound in trachytic rocks in which feldspar is more abundant than pyroxene. I believe, however, that a dolerite which I found in the road from Ovexeras to the hot springs of Comangillo, near Guanaxuato, belongs to the basalts of Caldera and Aguas buenas, and not to real trachyte. There is the same uncertainty on the position of phonolite, when it occurs insulated, or distant from basaltic and trachytic mountains. This insulated position characterises the phonolite of Peñon, which forms a bank in the Rio Magdalena, and appears immediately superposed on the granite of Banco; the phonolite which I saw piercing through the rock-salt of Huaura (Lower Peru, near the coast of the south sea); finally, that which rises at the northern banks of the steppes of Calabozo (Cerro de Flores). The latter is geognostically connected with pyroxenic amygdaloid alternating with transition greenstone. (Humboldt, *Rel. hist.*, t. i. p. 154.) The cellular amygdaloids (tezontli), containing glassy feldspar, pyroxene, and lithomarge, are most common in the central table-land of New Spain. They are sometimes covered by basalt, and sometimes form (Cuesta de Capulalpan) balls three feet thick, joined into cones, or hemispherical hills, superposed on transition porphyries.

III. Lavas that have issued in the form of currents from a crater. *Lithoid feldspathic lavas, similar to trachytes. Basaltic lavas. Obsidian of lavas. Vitreous pumice of obsidian.* We have already observed above, how rare are currents of lava in the Cordilleras. Those which I saw were owing to the lateral eruptions of Antisana, Popocatepetl, and Jorullo. Many currents (*Malpais*) have issued from volcanic craters which have since been closed, and which it is now impossible to discover. Other currents, having the same direction, are confounded with each other; they now appear in vast layers similar to more ancient pyroxenic rocks. Hornblende is much more rare in the lavas of the valley of Tenochtitlan (between San Augustin de las Cuevas and Coyoacan) than in the lavas of Europe. A learned Mexican mineralogist, M. Bustamante, has recently subjected them to a mechanical analysis with great success, according to the ingenious method suggested by M. Cordier. (*Semanario de Mexico*, 1820, No. 20. p. 80—90.)

IV. Volcanic tufa, often mixed with shells.

V. Local calcareous and gypseous formations, superposed on volcanic tufas, on basaltic rocks (mandelstein) or on trachytes. I include in these very modern formations in the table-land of Quito, the lamellar gypsum of Pululagua, the argillaceous and fibrous gypsum of Yaruquies, the slaty, carbureted, and vitriolic clays of San Antonio, the saliferous clays (?) of the Villa of Ibarra, the sand with lignites of Llano de Tapia (at the foot of Cerro del Altar), and the calcareous tufa (*caleras*) of Agua santa. In the Canary Islands also, calcareous, oolitic, and gypseous formations are subordinate to volcanic tufa. (Lancerote and Fortaventura). We cannot point out the relative age of these small deposites in comparing them with the chalk, or the most modern tertiary formations (§§ 37—39.); we have placed them here according to the order of their position above volcanic rocks. In Hungary, according to the interesting observations of M. Beudant, a sandstone with lignites (§ 35.) superposed on trachytic conglomerate (Dregely), on conglomerate of pumice (Palojta), and even on trachyte (Tokai), is covered, in its turn, either by calcaire grossier (§ 36.) of the tertiary formation, or by fresh-water limestone, or finally, by basaltic currents.

Such are the principal formations of pyrogenic rocks, owing to heaving up, to lateral overflowing, or to simple ejections. We shall confine ourselves to the description of facts, without touching on problems still so imperfectly known. We fear that what Montaigne said of a certain kind of philosophy, might be reasonably applied to geognosy, " It comes from our having inquisitive minds and bad eyes."

TABULAR ARRANGEMENT

OF

FORMATIONS

OBSERVED IN BOTH HEMISPHERES (1822).

Roman numerals are prefixed to the names of those formations, which, being very seldom wanting, and consequently extending most generally, may be considered as geognostic horizons. The sections and the pages where the descriptions are found are also given.

PRIMITIVE FORMATIONS.

The five latter formations, placed between gneiss and primitive mica-slate, are parallel formations.

TERTIARY FORMATIONS.

V. Fresh-water formation with porous millstones (meulière), above the sandstone of Fontainebleau (limestone with lymneæ), § 39. p. 403—404.

FORMATIONS (EXCLUSIVELY) VOLCANIC.

General views, p. 405—422.

I. Trachytic formations, p. 422—443.
Granitoïd and syenitic trachytes.
Porphyritic trachytes (feldspathic and pyroxenic).
Phonolites of trachytes (semi-vitreous trachytes).
Pearlstone with obsidian.
Millstone and cellular trachytes, with siliceous nodules.

Trachytic and pumice conglomerates, with alumstone, sulphur, opal, and opalised wood.

II. Basaltic formations, p. 442.
Basalt with olivine, pyroxene, and a little hornblende.
Phonolite of basalt.
Dolerite.
Cellular mandelstein.
Clay with pyrope-garnets.

This small formation seems to be connected with the clay with lignite of the tertiary formation, over which currents of basalt are often spread.

Conglomerates and basaltic scoriæ.

III. Lavas that have issued from a volcanic crater (ancient lavas, vast masses generally abounding in feldspar; modern lavas with distinct currents of small breadth; obsidian with lava and pumice of obsidian), p. 457.

IV. Volcanic tufa, with shells, p. 458.

Deposites of compact limestone, marl, clay with lignite, gypsum, and oolite, super posed on the most modern volcanic tufas. These small local formations belong perhaps to the tertiary rocks. Table-land of Riobambo; isles of Fortaventura and Lancerote.

In order to have more general ideas, and comprehend better the *relations of superposition* indicated in the table of rocks, *a pasigraphic method* may be employed, of which it will be useful to recapitulate here the fundamental principles. This method is double; it is either *figurative* (graphic, imitative) representing the superposed beds by parallelograms placed one above the other; or *algorithmic*, indicating the superposition of rocks, and the age of their formation, as the terms of a series.

I followed the first method in the *Tables de pasigrafia geognostica* which I traced, in 1804, for the use of the school of mines at Mexico; and which is generally designated by the name of *sections of formations*. It offers the advantage of addressing itself to the eye more directly, and of expressing *simultaneously in space* two series or systems of rocks, which cover the same formation. It furnishes an easy method of indicating the *geognostic equivalents*, or *parallel rocks*, as also in the case where, by the local suppression of the formation β, the formation α immediately supports γ. Two parallel rocks, for instance, clay-slate and quartz rock (page 120.), both superposed on primitive mica-slate, are represented in the figurative method by two parallelograms of the same height placed upon a third. The names of the rocks are inscribed in the parallelograms, or as we shall see below, they are characterised by covering them with a species of netting, differently modified, according as the rocks graphically represented pass, or do not pass to each other. By the local suppression of the sandstone of Nebra (variegated sandstone) and the limestone of Gottinguen (muschelkalk), Jura limestone may, in one place, repose immediately (pp. 357. 373.) on alpine limestone (zechstein), while, in another, we see following from below to above, alpine limestone, muschelkalk, variegated sandstone, and Jura limestone. These relations of position will be expressed in one ideal section in retrenching from the lower part of the parellelogram which represents Jura limestone, on one side only, a quadrilateral figure, representing the two formations of muschelkalk and variegated sandstone.

The second method, which proceeds by series, and may be called *algorithmic,* indicates the rocks not in an imitative manner, not by *figured extension,* but by a special *notation.* The whole geognosy of positions being a problem of *series,* or the simple or periodical succession of *certain terms,* the various superposed formations may be expressed by general characters, for instance, by the letters of the alphabet. These notations applied to different parts of natural philosophy [1] in which the juxtaposition of

[1] Before the great discovery of the pile of Volta, I had, in my work on the *Irritation of the nervous fibre,* indicated, by a particular notation, in which cases, in a chain of heterogeneous metals and humid interposed parts, the muscular excitement took place, and in which the galvanic current was stopped. The simple inspection of the series, and the respective position of the terms (elements of the pile) might lead one to imagine the result of the experiment. (Humboldt, *Versuche über die gereizte Muskel-und Nervenfaser,* t. i. p. 236.)

things are examined, are not merely fanciful. They have the great advantage, in positive geognosy, of fixing the attention on the most general relations of *relative position*, *alternation*, and the *suppression* of certain terms of the series. The more we make abstraction of the value of signs (of the composition and structure of rocks), the better we seize, by the conciseness of a language in some degree algebraic, the most complicated relations of position, and the periodical return of formations. The signs α, β, γ, will no longer represent granite, gneiss, and mica-slate; red sandstone, zechstein, and variegated sandstone; chalk, tertiary sandstone with lignite, and Parisian limestone; they will only be the terms of a series, simple abstractions of the mind. We are far from pretending that the geognost ought not to study, even in its closest relations, the mineralogical and chemical composition of rocks, the nature of their crystalline texture, and their masses; we only desire that abstraction should be made of these phenomena when there is a question only of their *succession and relative age*.

If the letters of the alphabet represent these superposed rocks, of two series,

$$\alpha, \beta, \gamma, \delta \ldots\ldots$$

$$\alpha, \alpha\beta, \beta, \beta\gamma, \gamma, \delta \ldots\ldots,$$

the first indicates the succession of simple and independent formations; granite, gneiss, mica-slate, clay-slate or muschelkalk, sandstone of Königstein (quadersandstein), Jura limestone, and green sandstone with lignite (below the chalk). The second indicates the alternation of *simple* with *complex* formations; granite, gneiss-granite, gneiss, gneiss-mica-slate, clay-slate (pp. 86. 88.); or, to give an example taken from the transition formation (p. 129.), limestone with orthoceratites, limestone alternating with slate, transition slate alone, slate and grauwacke, grauwacke alone, transition porphyry..... In the *complex* formations, that is, in those which present the periodical alternation of several beds, we sometimes distinguish three different rocks, which do not pass to each other in the same group;

$$\text{or } \begin{matrix} \alpha,\ \beta,\ \alpha\beta\gamma,\ \gamma\ \ldots. \\ \alpha\beta\gamma,\ \alpha\beta\delta,\ \beta\alpha\varepsilon\ \ldots.., \end{matrix}$$

as in primitive formations, there are alternating beds of granite, gneiss, and mica-slate; so in the transition formation, alternating beds of grauwacke, slate, and limestone, or grauwacke, slate, and porphyry, or clay-slate, grauwacke, and greenstone, constitute one formation. In the transition rocks, as we have stated above, clay-slate and grauwacke alone are not the terms of the series. Those terms are all complex; they are groups, and grauwacke belongs at once to several of those groups. It thence results that the term *grauwacke formation* relates only to the predominance of that rock in its association with other rocks.

Each class affords the example of independent formations which *prelude* as subordinate beds. If

$\alpha\beta\gamma$, or $\alpha\beta$, $\beta\gamma$ indicate the complex formations of granite, gneiss, and mica-slate, or of granite and gneiss, clay-slate and porphyry, porphyry and syenite, marl and gypsum, that is, the formations in which beds of two, and even three rocks alternate indefinitely; $\alpha+\beta$, $\beta+\gamma$, will indicate that gneiss forms simply a bed in granite, porphyry in slate, &c. Then

$$\alpha,\ \alpha+\beta,\ \beta,\ \beta+\gamma,\ \gamma\ \ldots.$$

expresses the curious phenomenon of formations which *prelude* or announce themselves as subordinate beds. Those beds sometimes call to mind the terms that precede (*lower rocks*) sometimes the terms that follow (*upper rocks*). Thus we shall have

$$\alpha,\ \beta,\ \beta+\alpha,\ \beta,\ \beta\times\gamma,\ \gamma\ \ldots.$$

The porphyries and granular syenites of the transition formation penetrate into the red sandstone, forming subordinate beds. If the position of the formation of the valley of Fassa is such as has been lately announced (p. 340.), a preceding term (syenite) passes as far as into the alpine limestone or zechstein; it is the case in the series,

$$\alpha,\ \beta+\alpha,\ \gamma+\alpha,\ \delta\ \ldots.$$

When we would apply the pasigraphic notation also to the elements of composed rocks, that notation may indicate, how by the progressive augmentation of the elements of the mass, and chiefly by the insulated crystals, beds are formed by a sort of *interior developement;*

$$abc,\ abc^2,\ abc^3,\ \ldots.\ abc+c$$

We have preferred in this particular case (beds of feldspar in granite, beds of quartz in mica-slate or gneiss, beds of hornblende in syenite, beds of pyroxene in transition dolerite) the letters of the Roman alphabet to those of the Greek, in order not to confound the elements of a rock (feldspar, quartz, mica, hornblende, pyroxene) with the rocks that enter into the composition of complex formations.

We have hitherto shown that in making abstraction altogether of the composition and physical properties of rocks, the *pasigraphic notation* can reduce the most complicated problems of composition to great simplicity. This notation shows how the same subordinate beds (rock-salt in zechstein and red marl, §§ 28, 29.; coal in red sandstone, zechstein, and muschelkalk) pass across several formations, superposed on each other:

$$\alpha+\mu,\ \beta+\mu,\ \gamma,\ \delta+\mu\ \ldots\ldots$$

It also reminds us of the return of feldspathic and crystalline formations in the transition rocks and red sandstone (Norway, Scotland), a return which is analogous to that of granite after gneiss and primitive mica-slate:

$$\alpha,\ \beta,\ \alpha,\ \gamma,\ \delta\ \ldots\ldots\ \varkappa,\ \lambda,\ \alpha,\ \beta,\ \ldots$$

The first terms of the series re-appear, even after a long interval, after grauwacke and limestone with orthoceratites, that is, after *fragmentary* and *shelly* rocks.

In concluding this work, I shall show, that if we give less generality to the notation, and modify it according to some physical considerations (of structure and composition), we may, by means of twelve geognostic signs, exhibit the most important phenomena of position of the primitive, intermediary, secondary, and tertiary formations. Those twelve signs comprehend seven series of rocks; viz. mica-slate (and its modifications on one side into granite and gneiss, on the other, into clay-slate), the euphotides, the hornblende rocks (greenstone, syenite), porphyries, limestone, and fragmentary rocks. Characters have been added for the great deposites of coal and rock-salt, which will serve to guide geognosts, their position indicating that of red sandstone and alpine limestone.

Table and Value of the Signs.

α, Granite.
β, Gneiss.
γ, Mica-slate.
δ, Clay-slate.

The first four letters of the alphabet have been employed to designate the four most ancient primitive formations. As these formations pass gradually into each other, the letters have been chosen which succeed immediately in an alphabetical order. Granite passes to gneiss, gneiss to mica-slate, and this to clay-slate. Other formations (porphyry, greenstone, euphotide) appear in some degree insulated, often as if *superadded* to more ancient formations; they have, therefore, been represented by letters which do not immediately succeed each other, and do not follow the letters $\alpha, \beta, \gamma, \delta$. By these means the formations which are connected less with others than they are connected together (euphotide and greenstone), are distinguished in pasigraphic writing in as marked a manner as in nature.

ο, Ophiolite, euphotide, gabbro, and serpentine; every formation generally abounding in diallage.

σ, Syenite, greenstone; in general every formation abounding in hornblende.

π, Porphyry. We sometimes see π pass to σ, and σ pass to ο.

τ, Calcareous and gypseous formations (τιτανος). If we would individualise calcareous formations still more, we might distinguish the primitive (τ), and those which contain organic remains (τ'); we might even by exponents, indicate separately transition limestone (τ^t), alpine limestone or zechstein (τ^a), limestone of Gottinguen or muschelkalk (τ^m), Jura limestone or the great oolitic formation (τ^o), chalk (τ^c), coarse Parisian limestone (τ^p), &c.

κ, Fragmentary, arenaceous, and aggregated rocks, conglomerates, grauwacke, sandstone, breccia, clastic rocks of M. Brongniart (κλασμα).

The accentuation (κ') indicates, as in τ, that the sandstone is shelly. We may distinguish grauwacke, or the fragmentary transition rocks (κ^g); red sandstone (κ^a), containing the great deposite of coal (*anthrax*); variegated sandstone or sandstone of Nebra ($\varkappa^n$); sandstone of Königstein, or quadersandstein ($\varkappa^q$); green sandstone, or tertiary sandstone with lignites below the chalk ($\varkappa^l$); sandstone abounding more in lignite above the chalk ($\varkappa^{2l}$); sandstone of Fontainebleau ($\varkappa^f$), &c. A good notation should have the advantage of modifying the value of the signs according as we stop at divisions variously graduated. The exponents have an allusion to the names of the rocks.

ξ, Coal, of which the greatest deposite is found at the beginning of the secondary formation; the same sign accentuated (ξ') indicates lignite, of which the great deposite is placed at the beginning of the tertiary formation, and which is sometimes shelly (ξυλον).

ϑ, Rock-salt, of which the principal formation is sometimes found in alpine limestone, sometimes in red marl or variegated sandstone. Not being able to employ the first letter of the Greek word ἁλς (which already indicates granite), I have made allusion to ϑαλασσα.

||, The former division of formations into primitive, intermediary, secondary, &c., is indicated by two perpendicular *bars*. When the geognostic series have very numerous terms, this sign appears like points of repose. The experienced geognost knows previously where the first transition rock, coal-sandstone, is placed. The accentuation of a character (δ', τ', $\varkappa'$) calls to mind, in general, that a rock contains remains of shells, and is not primitive.

The following are some examples of the employment of those twelve pasigraphic signs of rocks:

$$\alpha, \gamma+\pi, \delta\tau', \kappa', \pi, \sigma, \alpha.$$

The transition formation begins after $\gamma+\pi$ (mica-slate, with beds of primitive porphyry). It is nearly the succession of the formations of Norway (p. 137.). Then follows a complex formation of clay-slate and black limestone, with remains of shells, grauwacke, porphyry, syenite, and granite. The terms $\delta\tau'$ and κ', which precede π, σ, α, characterise those three rocks as transition rocks. In England, where the transition rocks furnish two very distinct calcareous formations (that of Dudley and Derbyshire), we see in succession:

$$\beta, \sigma\pi, \delta', \kappa^{g}, \tau', \kappa^{g}, \tau', \xi, \kappa^{a}, \tau^{a}, \kappa^{n}+\vartheta, \tau^{o}, \kappa^{l}, \tau^{c}, \kappa^{21} \ldots.$$

The transition formation begins with that of syenite and porphyry (Snowdon) placed on gneiss supposed to be primitive; then follow, a clay-slate with trilobites, the grauwacke of May-hill, the transition limestone of Longhope, the old red sandstone of Mitchel Dean, the mountain limestone of Derbyshire, the great coal formation, the new red conglomerate which represents the red sandstone, magnesiferous limestone, red marl with rock-salt, oolitic limestone, secondary sandstone with lignite (green sand), chalk, tertiary sandstone with lignite, or plastic clay, &c. The secondary formations on the continent, if they were all developed, succeed in the following manner:

$$\tau', \kappa^{g} \,||\, \pi\kappa^{a}+\xi, \tau^{a}+\vartheta, \kappa^{n}, \tau^{m}, \kappa^{q}, \tau^{o}, \kappa^{l}, \tau^{c}, \,||\, \kappa^{21} \ldots.$$

In comparing this type with that of England,

$$\xi,\ \varkappa^{a},\ \tau^{a},\ \varkappa^{n}+\vartheta,\ \tau^{o},\ \varkappa^{l},\ \tau^{c}\ldots\ldots$$

we see, that between the oolites (τ^{o}) and the red marl or sand of Nebra ($\varkappa^{n}$) there are two formations suppressed in England, viz., muschelkalk and quadersandstein; coal (ξ), rock-salt (ϑ) and oolite (τ^{o}) serve as terms of comparison like a geognostic horizon. But on the continent ξ and ϑ are connected with red sandstone and alpine limestone, while in England these deposites are rather connected with the transition rocks and red marl. Sometimes τ^{a} is subordinate (p. 298.), intercalated in $\varkappa^{a}$: these terms of the series (alpine limestone and red sandstone) form only one. The uncertainty of knowing whether a limestone is alpine (zechstein) or transition, arises generally from the suppression of the red sandstone and the deposite of coal which contains sandstone.

Of the two series,

$$\tau,\ \varkappa+\xi,\ \tau\ldots,$$

$$\tau,\ \varkappa,\ \tau\ldots,$$

the first alone shows the certainty that the last τ is alpine limestone. In the second series, the two limestones and the fragmentary rock which separates them may be of transition. The close connection of chalk with Jura limestone is evident, according to the alternation of beds (τ^{o}, $\varkappa^{l}$, τ^{c}, $\varkappa^{21}$), and according to the analogy of the sandstone with lignite below and above the chalk.

In order to unite the principal phenomena of the position of rocks in the primitive, intermediary, secondary, and tertiary formations, I propose the following series:

$\alpha,\ \alpha\beta,\ \beta+\pi,\ \beta\gamma,\ \gamma+\tau,\ \alpha,\ \gamma,\ \delta,\ \alpha,\ \beta,\ \delta,\ o \parallel \varkappa^{g},\ \tau',\ \delta\tau',\ \delta',\ \delta'+\pi,\ \gamma,\ \tau',\ \sigma\pi,\ \sigma+\alpha,\ \sigma\pi,\ o \parallel \pi\varkappa^{a}+\xi,\ \tau^{a}+\vartheta,\ \varkappa^{n},\ \tau^{m},\ \varkappa^{q},\ \tau^{o},\ \varkappa^{l},\ \tau^{c} \parallel \varkappa^{2l},\ \tau^{p}\ldots$

It would be useless to give the explanation of those characters; it will be seen by comparing them with the table of formations. I shall confine myself to fixing the attention of the reader on the accumulation of porphyries (π) on the limits of the transition and secondary formations, the position of euphotide formations (o), the great deposites of coal and lignite (ξ), and on the return (almost periodical) of feldspathic formations of transition granite, gneiss, and mica-slate (α, β, γ). As the notation I here present may be variously graduated by the manner in which the characters are accentuated, in uniting them as co-efficients in complex formations, or in adding exponents, I doubt whether the names of the rocks arranged by series at the side of each other would address itself as forcibly to the eye as the algorithmic notation.

In the figurative or graphic method, that which represents the formations by parallelograms superposed to each other, we might also indicate the relations of composition and structure by characters covering like a net the whole surface of the parallelograms. In lengthening the granular parts of granite, and dividing the parallelograms in beds somewhat thick, we obtain the character of gneiss. In rendering the foliated texture undulating, and in interrupting it by nodules (of quartz), the character of gneiss is changed into that of mica-slate. The syenite might be represented in the same manner, by the sign of granite, to which might be added black points (hornblende). These characters may pass from one to the other, like the rocks which they indicate. By using them in sections, I formed very detailed drawings of the valleys of Mexico and Totonilco, the vicinity of Guanaxuato, and the road from Cuernavaca to the South Sea; those drawings have the advantage of not requiring the use of colours. I shall not enter more into detail on the characters which may be employed. They may be variously modified; the conciseness of notation, and the spirit of the pasigraphic method are all that it is important to attend to.

NOTES.

§ 1. Léopold de Buch., *Géogn. Beobacht.*, tome I. page 16. 23.; Id., *Reise nach Norwegen*, II. p. 188.; Id., *Gilbert's Annalen*, 1820, April, p. 130. Leonhard, *Taschenbuch*, 1814, p. 17. Freiesleben, *Bemerkungen über den Harz*, I. p. 142. Leonhard, Kopp et Gærtner, *Propædeutik*, p. 159. Bonnard *Essai géogn. sur l'Erzgebirge*, p. 18. 48.; Id., *Aperçu géogn. des terrains*, p. 32. D'Aubuisson, *Traité de géogn.*, II. 12. Jameson, *Syst. of Miner.*, III. 107. Goldfuss et Bischof, *Beschreibung des Fichtelgebirges*, I. 145.; II. 38. Boué, *Géologie d'Ecosse*, p. 16. 348.; *Geol. Trans.* II. 158. *Edinb. Phil. Trans.* VII. 350. Beudant, *Voyage minér. et géol. en Hongrie*, III. 19. 27. Humboldt, *Essai sur la géogr. des plantes*, p. 122.; Id., *Relat. histor. de voy. aux rég. équin.* II. 100. 299. 507.

§ 2. Raumer, *Geb. von Nieder-Schlesien*, p. 10.

§ 3. Bonnard, *Erzgeb.*, p. 62. 118. Goldfuss, *Fichtelg.*, I. 145. 148. 172.; II. 32.

§ 4. Pusch, in Leonh., *Taschenb.*, 1812, p. 42. Raumer, *Fragm.*, p. 33. 36. 70. Bonnard, *Erzgeb.* p. 104. 121. Maincke et Keferstein, in Leonh., *Taschenb.*, 1820, p. 103.

§ 5. Buch., *Beob.*, I. 33.; Id., *Norw.*, I. 197. 358. II. 240.; Id., in *Mag. naturf. Freunde*, 1809, p. 46. D'Aubuisson, *Géogn.* II. 60—66.; II. 183. 187. Blöde, Leonh. *Taschenb.*, 1812, p. 17. Humboldt, *Nivell. géogn. des Andes*, *Recueil d'observ. astron.*, I. 310.

§ 6. Bonnard, *Erzgeb.*, p. 72. Humboldt, *Rel. hist.*, I. 556. II. 139.

§ 7. Goldfuss, *Fichtelgeb.*, I. 172—174. Bonnard, *Terrains*, p. 34. 40. 82. 66.; Id., *Roches*, p. 34. Humboldt, *Rel. hist.*, I. 610,; II. 142. 233. 491. 569. 715.

§ 8. Burckhardt, *Travels in Syria*, p. 142. D'Aubuisson, *Géogn.*, II. 19.

§ 9. Steffens, *Oryktoguosie*, I. 270. Boué, *Ecosse*, p. 55. Humboldt, *Rel. hist.*, II. 40.

§ 10. Beudant, *Hongrie*, II. 213. Bonnard, *Terrains*, p 79.

§ 11. Buch, *Géogn. Beob.*, I. 45. 51. 124. 257.; Id., *Norwegen*, I. 191. 209. 219.; Id., *Nat. Mag.*, 1809, p. 115. Cordier, *Journ. des mines*, XVI. 254. Bonnard, *Terrains*, p. 46. D'Aubuisson, *Géogn.*, II, 73—93.; Id., *Journal de Physique*, p. 1807, 402. Eschwege, *Journal von Brasilien*, II. 14. Freiesleben, *Géogn. Beytrag zur Kenntniss des Kupfersch.*, V. 257. Goldfuss, *Fichtelg.*, p. 9.

§ 12. Buch, *Norwegen*, I. 272. 413.

§ 13. Buch, *Géogn. Beobacht.*, I. 30.; Id., *Norwegen*, II. 27. 31. Raumer, *Géogn. Versuche*, p. 50.

§ 14. Freiesleben, *Harz*, II. 66. Bonnard, *Erzgeb.*, p. 109—133.

§ 15. Beudant, *Hongrie*, II. 84. III. 30. 40. Buch, *Norwegen*, II. 83. 87.; Id., *Mag. naturf. Fr.*, 1810, p. 147. Boué, *Ecosse*, p. 386.

§ 16. Eschwege, *Journ. von Brasilien*, I. 25. 34. 36. 38.

§ 17. Eschwege, *Bras.*, II. 241.

§ 18. Bonnard, *Terrains*, p. 56.

§ 19. Buch, in *Mag. nat. Fr.*, 1810, p. 137.; Id., *Géogn. Beob.*, I. 68. 71.; Id., *Norwegen*, I. 479. II. 29. 84. 87. 135. Esmark, in Pfaff, *Nord. Arch.*, III. 199. Saussure, *Voyages dans les Alpes*, § 1362. *Journ. de Phys.* XXXV. 298. Targieni Tozzetti, *Viaggi*, II. 433. Brocchi, *Bibl. ital.*, IX. 76. 356. Beudant, *Hongrie*, III. 49.

§ 20. Brochant, *Observ. géol. sur les terrains de transition de la Tarantaise*, p. 16. 19. 31. 33. 37. 39. 44. 50. 53.; Id., *Mémoire sur les gypses anciens*, p. 12—46. Buch, *Mag. nat. Fr.* 1809, p. 181.; Id., Leonhard's *Taschenb.*, 1811, p. 335. Raumer, *Fragmente*, p. 10. 24. D'Aubuisson, *Journ. des mines*, n°. 128. p. 161.

§ 21. Beudant, *Hongrie*, III. 96. 133. 199. Raumer, *Neider-Schlesien*, p. 72.

§ 22. Charpentier, *Description géogn. des Pyrénées* (manuscrit), §§ 35. 66. 89. 100. 105. 141—167.; Id., *Mém. sur le gisement des gypses de Bex*, *Naturw. Anzeiger der Schweiz. Gesellsch*, 1819, n°. 9. p. 65. Raumer, *Fragmente*, p. 10. 32. 74.; Id., *Versuche*, p. 41. Buch, *Norwegen*, II. 281.; Id., *Mag. nat. Fr.*, 1809, p. 175. Meinecke et Keferstein, *Taschenb.*, p. 63. Haussmann, *Nord. Beytr.*, II. 77. IV. 653.; Id., *Reise durch Scandinavien*, II. 239. Engelhardt, *Felsgebäude Russlands*, I. 37. Keferstein, *Teutschland geognostisch dargestellt*, I. 136. Eschwege, *Brasil.*, II. 258. Maclure, *Géol. des Etats-Unis*, p. 24. Brongniart, *Notice sur l'histoire géogn. du Cotentin*, p. 17.; Id., *Crustacés fossiles*, p. 46—63. Beudant, *Hongrie*, III. 76. 578. Saussure, *Alpes*, § 501. Wahlenberg, *Acta Soc. Upsal.*, VIII. p. 19. Link, *Urwelt*, p. 2. Castelazo, *de la riqueza de la Veta Biscaina* (Mexico, 1820), p. 9. Humboldt, *Essai polit. sur la Nouvelle-Espagne*, II. 534. 537. 519—526.

§§ 23 & 24. *Del Rio, la Gazeta de Mexico*, XI. 416. Humboldt, *Essai polit.*, II. 494. 521. 581. 583. Beudant, *Hongrie*, II. 157. III. 67—124. 148. Boué, *Ecosse*, p. 147. Burckhardt, *Travels in Syria*, 1822, p. 493. 567. Raumer, *Fragm.* p. 24—26. 37. 48. Haussmann, in Moll's *Neuem. Jahrb.*, I. 34. Buch, *Norw.*, I. 96—144.

§ 25. Boué, *Ecosse*, p. 94. 358. Palassou, *Supplément aux Mémoires pour servir à l'hist. nat. des Pyrénées*, p. 139—153. Brongniart, *sur les Ophiolithes*, p. 26. 46. 56. 59. 61.

§ 26. Beudant, *Hongrie*, II. 575—580. 584—594. III. 171. 184. 194. 204. *Geol. Trans.* IV. p. 9. *Annales des mines*, III. p. 45. et 568. Steffens, *Geogn. Aufsätze*, p. 11. Buch, *Beob.* I. p. 104. 157. Heim, *Geogn. Beytr. zur Kenntn. des Thüring. Waldes*, II. 5te Abth., 236. Conybeare and Philipps, *Geol. of England*, I. 298. 312. 324—370.

§ 27. Humboldt, *Géogr. des plantes*, p. 128.; Id., *Essai politique*, II. 589.

§ 28. Escher, in Leonh. *Taschenb.* 1804, p. 347.; Id., in *Neue Zürcher Zeitung*, 1821, n°. 60. p. 237. Uttinger, in Leonh. *Taschenb.*, 1819, p. 42. Keferstein, *Teutschland*, III. 259. 263. 273. 340. 372. 390. 407. Mohs, in Moll's *Ephem.* 1807, p. 161. Lupin, *ib.*, 1809, p. 359. Ramond, *Voy. au sommet du Mont-perdu*, p. 15. 26. Traill, *Geol. Trans.* III. 138. *Bibl. univ.* XIX. 38. Buckland, *on the structure of the Alps*, p. 9. Buch, *Géog. Beob.* I. 153—171. 194. 216. 256. Freiesleben, *Kupfersch.*, IV. 284. Tondi, in Lucas, *Table méth. des esp. min.*, II. 243. Haussmann, *Nord. Beytr.* IV. 88. *Jénaer. litter. Zeit.*, 1813, p. 100. Steffens, *Geogn Aufs*, p. 49. Beudant, *Hongrie*, III. 231—237. Conybeare and Philipps, *England*, I. 301. Marzari Pencati, *Cenni geologici*, p. 21. Breislak, *Sulla giacitura di alcune rocce porfiritiche e granitose*, p. 25—35.

§ 29. Conybeare and Philipps, *Engl.*, I. 61. 269. Freiesleben, *Kupfersch.*, I. 90—188. IV. 276—284.

§ 30. Freiesleben, *Kupfersch.*, I. 65. 89. IV. 295—317. Raumer, *Versuche*, p. 112—115.

§ 31. Haussmann, *Nord. Beytr.*, 1806, st. 1. p. 73. Freiesleben, *Kupfersch.*, I. 102—107. IV. 283. 293. Conybeare and Philipps, *Engl.*, I. 122. Raumer, *Nieder-Schlesien*, p. 121. 123. 153.

§ 32. Humboldt, *über die unterird. Gasarten*, p. 39. Karsten, *Min. Tab.*, p. 63—65. Buch, *Landek.*, p. 7.; Id., *Helvet. Alm.*, 1818, p. 42. Gilb. *Annalen*, 1806, st. 5. p. 35. Escher, *Naturw. Anzeiger der Schweiz. Ges.*, *Jahrg.*, IV. p. 29. Charbaut, *Mém sur la géologie des environs de Lons-le-Saunier*, p. 7. 9. 24. 27. Merian, *Beschaffenheit der Gebirgsbild. von Basel*, p. 23. 36. 46. 83.

§ 33. Conybeare and Philipps, *Engl.*, I. 127—164.

§ 34. Brongniart et Cuvier, *Descr. géol. des environs de Paris*, 1821, p. 10—17. 68—101. Steffens, *Geogn. Aufs.*, p. 121. Raumer, *Vers.*, p. 85. 116. Conybeare and Philipps, *Engl.*, I. 60–126.

§ 35. Bonnard, *Terrains*, p. 226. Brongniart, *Descr. géol.*, p. 17—28. 102—122. Conybeare and Philipps, *Engl.*, I. 37—57. Raumer, *Vers.*, p. 120—122. Beudant, *Hongrie*, III. 242—264. Lardy, in *Bibl. univ.*, March 1822, p. 180. 183. Keferstein, *Teutschland*, I. 46. Freiesleben, *Kupfersch.*, V. 255. Adolphe Brongniart, *Classific. des végétaux fossiles*, p. 54.

§ 36. Beudant, *Hongrie*, III. 264—282. Brongniart, *Descr. géol.*, p. 29—38. 123—203.

§ 37. Raumer, *Vers.*, p. 123—125. Brongniart, *Descr. géol.*, p. 38—50. 203—263.

§ 38. Raumer, *Vers.*, p. 125. D'Aubuisson, *Géognosie*, II. 414. 417. Brongniart, *Descr. géol.* p. 50—56. 264—274. Bonnard, *Terrains*, p. 217.

§ 39. Brongniart, *Descr. géol.*, p. 57—60. 275—320. Beudant, *Hongrie*, III. 282—283.

§ 40. Buch, *Géogn. Beob.*, II. 172—190. Id., in *Mag. nat. Fr.*, 1809, p. 299—303.; Id., in *Mém de Berlin.*, 1812, p. 129. —154. Fleuriau de Bellevue, *Journ. de phys.*, LI. et LX. Cordier, *Mém. sur les substances minérales, dites en masse, qui entrent dans la composition des roches volcaniques*, p. 17—69. *Bustamente sobre las lavas del Padregul de San Augustin de la Cuevas,* in *Seman. de Mexico*, 1820, p. 80. Leonhard, *Propædeutik*, p.168—175. Ramond, *Nivellement barométrique et géognostique de l'Auvergne*, p. 32—45. Breislal, *Introd. a la geologia*, I. 234. 261. 316. Heim, *Thüringer-Wald*, p. 229. Singer, in Karsten's *Archiv für Bergbaukunde*, III. 88. Robiquet, in *Annales de physique et de chimie*, XI. 206. Nose, *Niederrheinische Reise*, II. p. 428. Boué, *Ecosse*, p. 219—287. Beudant, *Hongrie*, III. 298—644. Humboldt, *Essai sur la géographie des plantes, et tableau physique des régions équinoxiales*, p. 129. Id., *Essai polit.*, I. 249—254.; Id., *Nivellem géogn. des Cordillères*, in *Recueil d'obs. astron.*, I. 309—311. 327. 332.; Id., *Recueil d'obs. de zool. et d'anat. comparée*, I. 21.; Id., *Relat. hist.*, I. 91. 116. 129. 133. 136. 148. 151. 153—155. 171. 176. 180. 308. 312. 394. 640.; II. 4. 14. 16. 20. 25. 27. 39. 452. 515. 565. 719.

THE END.

London:
Printed by A. & R. Spottiswoode,
New-Street-Square.

Printed in the United States
By Bookmasters